Sparsity

Sparsity

Its practical application to systems analysis

A. Brameller

R. N. Allan

Department of Electrical Engineering
University of Manchester
Institute of Science and Technology

and

Y. M. Hamam

American University, Beirut

Pitman Publishing

First published 1976

Pitman Ltd
Pitman House, Parker Street, Kingsway, London WC2B 5PB
PO Box 46038, Banda Street, Nairobi, Kenya

Pitman Publishing Pty Ltd
Pitman House, 158 Bouverie Street, Carlton, Victoria 3053, Australia

Pitman Publishing Corporation
6 East 43 Street, New York, NY 10017, USA

Sir Isaac Pitman (Canada) Ltd
495 Wellington Street West, Toronto 135, Canada

The Copp Clark Publishing Company
517 Wellington Street West, Toronto 135, Canada

ISBN 0 273 00938 9

Text set in 10/12 pt. IBM Press Roman, printed by photolithography, and bound in Great Britain at The Pitman Press, Bath.

T2:98

Preface

Engineers and scientists have always been and will continue to be vitally concerned with the development and solution of analytical models which realistically describe physical systems. Model building and analysis has been successfully applied in all areas of physical science and to problems of economics; social, medical and management science; logistics and environmental analysis. The areas of application are literally too numerous to enumerate. The two activities, model building and model analysis, are complementary and together form part of the general heading of system analysis. It provides a valuable appreciation of the behaviour of physical systems when subjected to various stimuli and the sensitivity of this behaviour to parameter variation. From a practical point of view, the extent to which system analysis can be conducted is generally inhibited by inability to obtain an economic solution for the developed model. The most dramatic aid to the development of system analysis was the digital computer and the fact that there is now almost universal availability of large high speed digital systems. System analysis has become a relatively conventional and accepted form of investigation. Its application is not confined to the domain of the mathematician, the numerical analyst and the systems engineer, but includes the actual practitioners of many disciplines. Initial computer applications simply used the tremendous power of the computer to free the analyst from the repetitive and boring calculations previously done by hand or with mechanical calculators. The tremendous numerical potential of matrix techniques, which previously involved too much labour to be practical in large scale problems, was quickly realized and a new era in systems analysis began. Techniques were developed which were based on and exploited the tremendous iterative ability and precision of the modern high speed computer. This phase is now virtually complete and model analysis is now basically inhibited by saturation of the available computing power either by excessive computation times or storage requirements. It is therefore of considerable practical importance to augment the efficiency of computation and storage utilization by establishing logical and sequential techniques which can be readily appreciated by the practitioner and do not require considerable experience or the superior insight normally associated with the professional systems analyst. This is essentially the basic objective of this book. It presents in a readable manner the concept of sparsity and its application in system analysis. It also provides a useful application – orientated review of numerical techniques in the solution of systems of simultaneous linear equations. The book is not an elegant treatise on mathematical rigor but it is a sequential development of computational techniques with the dominant theme of exploiting the concept of sparsity to its fullest. The authors' backgrounds are in the electrical engineering field. The concepts contained in the book are not restricted to electrical engineering or even to engineering in general, but are applicable to the numerical solution of all physical systems modelled by large sets of simultaneous linear equations.

Roy Billinton
Professor, Power Systems Research Group
Department of Electrical Engineering
University of Saskatchewan
Saskatoon, Canada

Acknowledgements

We would like to express our gratitude to Professor L. M. Wedepohl, formerly Chairman of Electrical Engineering and Electronics, UMIST, and presently Dean of the Faculty of Engineering, University of Manitoba, Canada, for the deep interest and encouragement he showed during the preparation of this book, to Professor Roy Billinton, Department of Electrical Engineering, University of Saskatchewan, Canada, for his enthusiasm and for agreeing to write the Preface, to Dr K. Zollenkopf who pioneered the bi-factorization method of solving sparse, linear equations, to our colleagues in the Department of Electrical Engineering and Electronics, UMIST, who have given advice and help in the writing of this book, and to Mrs Dora Chantler for her patience and precision in the typing of the original manuscript. Finally the book would not have been possible without the immensely important foundations laid down by eminent mathematicians such as Gauss, Newton, Crout, etc., and many others whose names appear in the references.

Contents

Preface v

Acknowledgements vii

1 Introduction 1

2 Graphs and the formulation of linear network equations 4

2.1 Introduction 4
2.2 Network incidence matrices 5
2.2.1 Graphs and network terminology 5
2.2.2 Branch-nodal incidence matrix 6
2.2.3 Tree and co-tree incidence matrices 7
2.2.4 Branch-path incidence matrix 8
2.2.5 Branch-loop incidence matrix 9
2.3 Nodal formulation of linear network equations 11
2.3.1 Formulation of equations 11
2.3.2 Properties of the nodal coefficient matrix 14
2.4 Loop formulation of linear network equations 16
2.4.1 Formulation of equations 16
2.4.2 Properties of the loop coefficient matrix 18
2.5 Transformation of nodal flows to loop potential sources 19
2.6 Sparsity of the network equations 21
2.6.1 Sparsity coefficient 21
2.6.2 Sparsity of the nodal coefficient matrix 22
2.6.3 Sparsity of the loop coefficient matrix 23
2.7 Coefficient matrices with incidence symmetry 25

3 Solution of Simultaneous Linear Equations 27

3.1 Introduction 27
3.2 Direct Methods 28
3.2.1 Gauss elimination 28
3.2.2 Crout elimination 32
3.2.3 Matrix inversion 35

3.2.4 Pivoting 39
3.3 Asymmetry and Changes in a Network 40
3.4 Iterative Methods 47
3.4.1 Gauss–Seidel iteration 47
3.4.2 Relaxation method 51
3.4.3 Acceleration 53
3.5 Ill-conditioning 53
3.6 Choice of Method 55
3.6.1 Factors affecting choice 55
3.6.2 Comparison of methods 56

4 Matrix Factorization 58
4.1 Introduction 58
4.2 Product form of the inverse 59
4.3 Triangulation of matrices 63
4.3.1 LH factorization 64
4.3.2 LDH factorization 69
4.4 Bi-factorization 71
4.5 Comparison between triangulation and bi-factorization methods 77

5 Sparsity-directed Elimination 80
5.1 Introduction 80
5.2 Matrix elimination and graph reduction 80
5.3 Principles of ordering 83
5.4 Pre-ordering techniques 86
5.4.1 Least number of connected branches 86
5.4.2 Diagonal banding 87
5.5 Dynamic ordering techniques 91
5.5.1 Least number of connected branches 91
5.5.2 Introduction of least number of new branches 93
5.6 Comparison of ordering schemes 94
5.7 Network decomposition 96

6 Sparsity-directed Programming 100
6.1 Introduction 100
6.2 Storage of a list of numbers 101
6.3 Storage of sparse matrices 104
6.3.1 Scheme I 104
6.3.2 Scheme II 105
6.3.3 Scheme III 107
6.4 Programming principles for ordering and factorization 110
6.5 Programming principles of transformation matrices 119
6.6 Principles of program organization 124
6.7 Computational analysis of typical systems 125

7 Application of Sparsity Techniques to Linear Programming Problems 133
7.1 Introduction 133

7.2 Application to the revised simplex method 133
7.2.1 Basic feasible solution 133
7.2.2 Improving a basic feasible solution 134
7.2.3 Interchanging basic and non-basic variables 134
7.2.4 Steps of solution 136
7.2.5 Storage scheme 136
7.2.6 Numerical example 138
7.3 Application to trans-shipment problems 142
7.3.1 The trans-shipment problem 142
7.3.2 Basic feasible solution 145
7.3.3 Improving a basic feasible solution 147
7.3.4 Interchanging basic and non-basic variables 149
7.3.5 Steps of solution 151
7.3.6 Storage scheme and computational aspects 152
7.4 Other linear programming problems 156

8 Application of Sparsity Techniques to Non-linear Problems **158**
8.1 Introduction 158
8.2 Application to the load flow problem 159
8.2.1 Basic load flow equations 159
8.2.2 Solution of load flow problem 160
8.3 Application to general fluid flow problems 161
8.3.1 General method of solution 161
8.3.2 The flow equations 162
8.3.3 Formulation of loop equations 163
8.3.4 Nodal solution of Jacobian equations 165
8.4 Application to non-linear optimization problems 168
8.4.1 Non-linear optimization in power system analysis 168
8.4.2 Formulation of the optimization problem 168
8.4.3 Solution of the optimization problem 169
8.5 Concluding remarks 172

Index **174**

1
Introduction

Many aspects of engineering and non-engineering are frequently concerned with the analysis of problems which are defined by a set of simultaneous linear equations and are sparse, that is, they have a large number of zero coefficients. In general, these equations can be mathematically modelled as a network and solved by the network approach. Until fairly recently the size of the problem and the scope of the analysis had to be restricted to prevent excessive computation times and storage requirements. Using modern techniques it is now possible to solve very quickly and efficiently a set of simultaneous equations having up to, perhaps, several thousand variables. It is therefore possible to analyse much larger and more complex types of problem in greater detail. These methods permit engineers and non-engineers involved in such analyses to tackle problems of vital and much greater practical significance, since, in reality, their systems, networks or physical problems can be very large and complex.

The problems to which these modern techniques can be applied are many and varied but they usually fall within the general categories of design, operation and/or optimization. These general types of problems may be related to a physical network such as electricity supply, gas distribution, water distribution, communication systems, mechanical and civil engineering structures, etc., or to a physical problem such as traffic flow, urban planning, transshipment, cash flow, control problems, industrial organization and communication structures, genetic problems, behavioural and social science problems, etc. It is clear, therefore, that the source of the problem can be considerably different. The type of problem can be very similar, however, and can be solved using similar techniques based on the network approach.

In some cases these problems may be formulated directly as a set of simultaneous linear equations, in which case the network approach can be applied to the complete problem. In other cases, however, the simultaneous linear equations may be created as a subset of a much larger problem, as occurs, for example, in the successive linearization of a non-linear problem or in the interactive process of a computer-aided design project. It is in these interactive and iterative types of problems that the modern techniques for solving the network type of problem are of immense importance, since, as very efficient subroutines of an overall problem, they permit analyses that were previously impossible or only carried out in an approximate or oversimplified manner.

Because present-day problems of practical importance are becoming increasingly complex in character, their formulation and solution are becoming more difficult and much use is now made of digital computers in their solution. This requires that the mathematical formulation and solution of the problem is accomplished in a logical manner, and a great deal of care and skilful programming is essential to ensure efficient use of the computer.

The simultaneous linear equations describing the behaviour of a given network or network-type problem are derived from basic principles using, for example, Ohm's Law or Hooke's Law, or derived as a subset of a larger overall problem. After establishing these equations, consideration must be given to the most suitable method of solution. This should take into account the simplicity of the method, the number of arithmetical operations and the computer storage requirements. There are many numerical methods for solving simultaneous linear equations and the choice depends on the nature of the problem and the resources available. These methods can be simply the direct application of matrix algebra. Although much vital work can be and has been accomplished by these methods, however, they involve the direct manipulation of matrices without attempting to recognize whether the matrices have any special properties that could make the solution more efficient. Consequently it is possible that excessive computation time and storage space is used. In days when cost-effectiveness in all engineering and non-engineering applications is considered essential, this inefficiency should and often can be avoided.

One particular property associated with many problems of the network-type is known as sparsity, that is, the coefficient matrix associated with the problem contains a large proportion of zero elements. Physically this means that only a small number of branches of the network or graph are connected to each node or junction. In an electric-power system, for example, the ratio between the number of branches and number of nodes is about 1.5, and in a low-pressure gas-distribution system, the ratio is about 1.3. Taking an electric-power system as an example, therefore, the number of non-zero elements in the coefficient matrix of a 250 node problem is about one per cent, and in the coefficient matrix of a 2500 node problem is about 0·1 per cent. It is evident that if only the non-zero elements are stored and processed, a considerable increase in computational efficiency would be gained. The application of sparsity techniques and programming permits this increase to be achieved.

The purpose of this book is to encourage the exploitation of sparsity and to explain the numerical and computational programming techniques that can be used to solve problems possessing a significant degree of sparsity. The book limits itself to linear network or network-type equations because generally the problem either falls into this category or, if non-linear, is approximated to such a problem by successive linearization using iterative methods. Before the specific application of sparsity techniques can be discussed, it is essential to introduce and summarize some of the general aspects of linear networks, the formulation of their equations and the various methods that can be used to solve them. These aspects are covered in the subsequent chapters of this book and should permit a reader to understand fully how the exploitation of the sparsity structure enables the solution of a linear problem to be achieved much more efficiently.

All the numerical and programming techniques that are introduced and discussed in the book are illustrated by simple but complete numerical examples. These should allow a reader to feel thoroughly at ease with all the techniques and enable him to apply these techniques to his own particular problems.

The book does not contain any rigorous or detailed mathematical treatises, since it is basically a book on numerical techniques. Therefore the level of mathematics required to understand the book is quite minimal, although familiarity with matrices and matrix algebra is desirable. On the other hand, it is essential for the reader to be thoroughly familiar with the mathematical treatment of his own field of interest, in order to develop the techniques for solving his own problems. As discussed previously, the diversity of the problem is so large that it is impossible to describe all applications of sparsity techniques and programming. The two chapters in this book on applications are therefore intended to be representative of those problems that frequently occur in practice. The authors consider that readers, unfamiliar with these problems or the relevant mathematics, will benefit considerably from the logical techniques used to solve the problems.

The book itself is directed at any reader who is involved in the analysis of a problem that is either defined directly by a set of simultaneous linear equations or which arises during the solution of a much larger problem. In particular, the authors consider that engineers and non-engineers working in industry, commerce, local government, the civil service and similar organizations will improve the efficiency and cost-effectiveness of their computing demands and also increase the practical importance of their analyses if they incorporate the techniques and programming discussed in this book. In addition, the book forms a very suitable text for postgraduate and advanced undergraduate teaching, either as a formal textbook or in a programmed learning course.

2
Graphs and the Formulation of Linear Network Equations

2.1 Introduction

The solution of a given linear network problem requires firstly the formulation of a set of equations describing the response of the network and secondly the manipulation of the coefficient matrix so produced. Several methods exist for formulating these equations, these methods being based on either nodal or loop techniques. The associated coefficient matrix will differ depending on the techniques chosen, because these affect the number of equations and the number of non-zero coefficients in them. Therefore the method for solving the problem and the ability to exploit sparsity techniques most efficiently depend upon this formulation.

It is often found that the equations describing a given system are non-linear and therefore linear solution methods cannot be applied directly. Most of the methods for the numerical solution of non-linear equations, however, involve iterative processes such as the Newton–Raphson method. These methods involve successive linearization of the problem, resulting in a series of approximations of increasing accuracy. Because each approximation in the iterative process represents linearization of the problem to some degree of accuracy, each step can be represented by a hypothetical equivalent linear network. In the solution of a set of non-linear equations, the loop formulation may sometimes produce better conditioned equations, resulting in a smaller number of iterations. However, to take advantage of sparsity, in order to reduce computer storage space and the solution time, it may be more convenient to transform the linearized equations at each iterative step to nodal form. These successive transformations can be made from a knowledge of the network layout alone. Therefore, even for the solution of non-linear equations, it is useful to have a clear understanding of the concept of formulating linear network equations using incidence matrices which are created from the physical layout of the network.

Many important industrial problems are concerned with the economic distribution of a certain commodity. These are known as trans-shipment problems and are special cases of general linear programming. Trans-shipment problems have a special nature which makes it possible to solve them by the network approach. This nature is discussed in detail in chapter 7. Since a problem of this sort can be solved using the network approach, a knowledge of graph theory, as presented in this chapter, provides a better understanding of how a more efficient solution to the problem can be obtained systematically and logically and how sparsity can be exploited.

This chapter is therefore intended to introduce important standard terminology, to define the various incidence matrices and to compare the differences between the available methods of formulating the problem. This is an important prerequisite for understanding how sparsity is produced and evolved and how its property can be exploited most efficiently in solving all types of linear and linearized network problems.

2.2 Network incidence matrices

2.2.1 Graphs and network terminology

A network is an assembly of interconnected *branches*. The geometrical layout of the network, referred to as the *topology*, can be described by an equivalent *graph*. The same graph can represent a number of physical problems, such as the layout of an electrical network, a gas network or a mechanical structure. The graph shown in Fig. 2.1(*a*), for example, represents both the electrical network and mechanical structure shown in Fig. 2.1(*b*) and (*c*). Therefore networks of different physical types but with the same nodal and branch layout are analogous and mathematically they can be treated identically, although the fundamental equations describing the problem may be different.

A graph in which it is possible to trace a path between any two of its nodes is a *connected graph*. A disconnected graph is little more than a mathematical concept and since it may be considered to consist of several connected graphs, only connected graphs will be discussed. A *sub-graph* is obtained from the

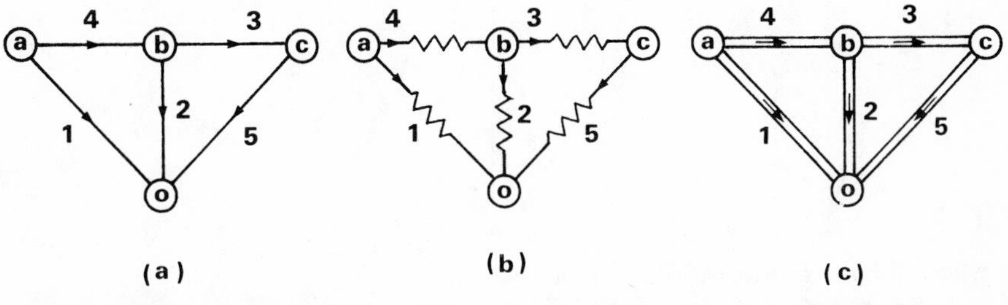

Fig. 2.1 Geometrical representation of physical problems (*a*) graph; (*b*) electrical network; (*c*) mechanical structure

original graph by removing some of the branches. The sub-graph may be connected or disconnected, depending on the branches removed.

In order to define the graph it is necessary to assign a direction to each branch which need not be the same as the flow or *through* quantity in the branch. As both the flow and branch directions are initially assigned arbitrarily, however, it is convenient to assume the direction of each branch and its flow quantity to be the same. A graph in which a direction is assigned to the branches is a *directed graph,* as shown by the example in Fig. 2.1(*a*).

2.2.2 Branch-nodal incidence matrix

A graph as shown in Fig. 2.1(*a*) may be described in terms of a *connection* or *incidence* matrix. Such a matrix may be constructed from a list of the branches with their terminal nodes as shown in Table 2.1.

Table 2.1 – Branch list for Fig. 2.1(a)

branch number	1	2	3	4	5
sending node	*a*	*b*	*b*	*a*	*c*
receiving node	*0*	*0*	*c*	*b*	*0*

The incidence matrix is rectangular, with the number of rows equal to the number of nodes and the number of columns equal to the number of branches in the network. Therefore the element c_{ij} in row i and column j of the incidence matrix $\mathbf{C}'$ corresponds to node i and branch j. It is defined as:

$$c_{ij} = \begin{cases} +1, & \text{if branch } j \text{ is directed away from node } i \\ -1, & \text{if branch } j \text{ is directed towards node } i \\ 0, & \text{if branch } j \text{ is not connected to node } i \end{cases}$$

This matrix is called the *non-reduced branch-nodal* incidence matrix and for the example shown in Fig. 2.1 is:

$$\mathbf{C}' = \begin{array}{c} \\ o \\ a \\ b \\ c \end{array} \begin{array}{c} \begin{array}{ccccc} 1 & 2 & 3 & 4 & 5 \end{array} \\ \begin{bmatrix} -1 & -1 & \cdot & \cdot & -1 \\ 1 & \cdot & \cdot & 1 & \cdot \\ \cdot & 1 & 1 & -1 & \cdot \\ \cdot & \cdot & -1 & \cdot & 1 \end{bmatrix} \end{array}$$

where the dots indicate zero elements.

For network applications it is necessary to select a reference node – mathematically this is referred to as the dependent node, since all branch and nodal quantities are dependent on it. A new incidence matrix can be created which does not include the row corresponding to the reference node. This new matrix is defined as the *reduced branch-nodal* incidence matrix. As will be seen more clearly later, the reduced matrix is of more interest and will therefore be referred to as the branch-nodal incidence matrix. For the network shown in Fig. 2.1 with node o chosen as the reference node, the branch-nodal incidence matrix is:

$$\begin{array}{r} \\ a \\ \mathbf{C} = b \\ c \end{array} \begin{array}{c} \begin{array}{ccccc} 1 & 2 & 3 & 4 & 5 \end{array} \\ \left[\begin{array}{ccccc} 1 & \cdot & \cdot & 1 & \cdot \\ \cdot & 1 & 1 & -1 & \cdot \\ \cdot & \cdot & -1 & \cdot & 1 \end{array}\right] \end{array}$$

2.2.3 Tree and co-tree incidence matrices

A connected sub-graph whose number of branches is one less than the total number of nodes is defined as a *tree* of the graph. The branches removed in order to obtain the tree are defined as a *co-tree* of the graph. Although a tree must always be a connected sub-graph, the co-tree may be connected or disconnected.

For any given graph a finite number of different trees can be traced. For the network shown in Fig. 2.1, a tree can be constructed as shown in Fig. 2.2. In fact for this example there is a total of eight different trees than can be traced as shown in Fig. 2.3.

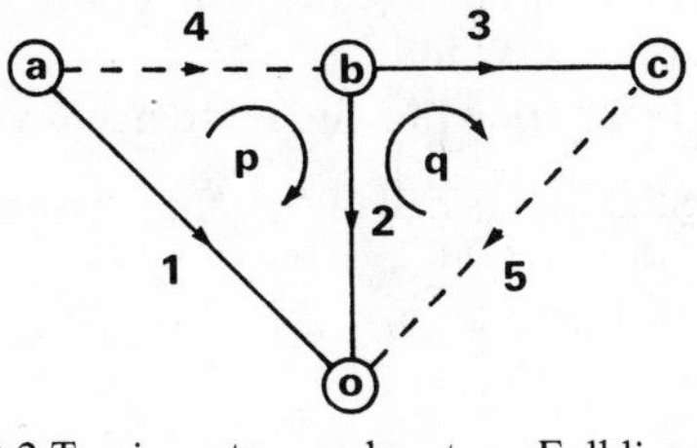

Fig. 2.2 Tracing a tree and co-tree. Full lines indicate tree branches and dotted lines indicate co-tree branches

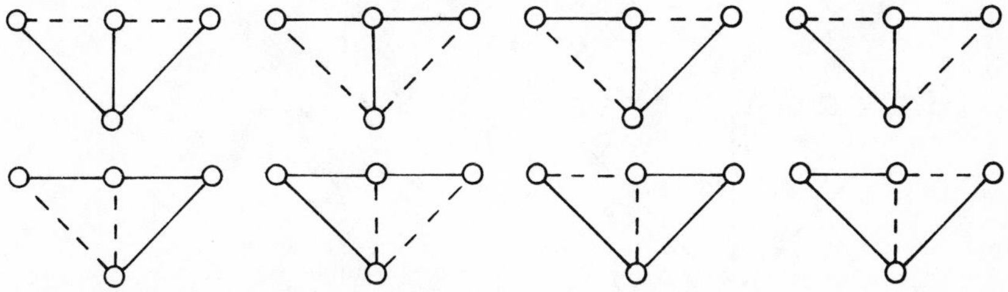

Fig. 2.3 Alternative trees and co-trees

By rearranging, if necessary, the numbering order of the branches, the branch-nodal incidence matrix can be partitioned to give the tree and co-tree incidence matrices. For the example shown in Fig. 2.2, the incidence matrix is partitioned as follows:

$$\mathbf{C} = \begin{array}{c} a \\ b \\ c \end{array} \overset{\begin{array}{ccccc} 1 & 2 & 3 & 4 & 5 \end{array}}{\left[\begin{array}{ccc|cc} 1 & \cdot & \cdot & 1 & \cdot \\ \cdot & 1 & 1 & -1 & \cdot \\ \cdot & \cdot & -1 & \cdot & 1 \end{array}\right]} = \left[\mathbf{C}_t \mid \mathbf{C}_c\right]$$

tree co-tree

The partitioned matrix $\mathbf{C}_t$, corresponding to the tree sub-graph of the branch-nodal incidence matrix, is square and non-singular. However, because the rows of the non-reduced branch-nodal incidence matrix are linearly dependent, due to the linear dependence of the equations, the sum of the elements in every column is zero and therefore any vertically partitioned square matrix created from this matrix is singular.

2.2.4 Branch-path incidence matrix

When a tree is defined for a given graph, a unique path can be traced from any node to the reference node, using the tree branches. An incidence matrix **A** describing these paths can be constructed with the element a_{ij} in row i and column j defined as:

$$a_{ij} = \begin{cases} +1, & \text{if the direction of branch } j \text{ is the same as the direction of a path from node } i \text{ to the reference node} \\ -1, & \text{if the direction of branch } j \text{ is opposite to the direction of a path from node } i \text{ to the reference node} \\ 0, & \text{if branch } j \text{ is not a path from node } i \text{ to the reference node} \end{cases}$$

This matrix is again square and non-singular and is called the *branch-path* incidence matrix.

Consider the tree shown by full lines in Fig. 2.4. For this example

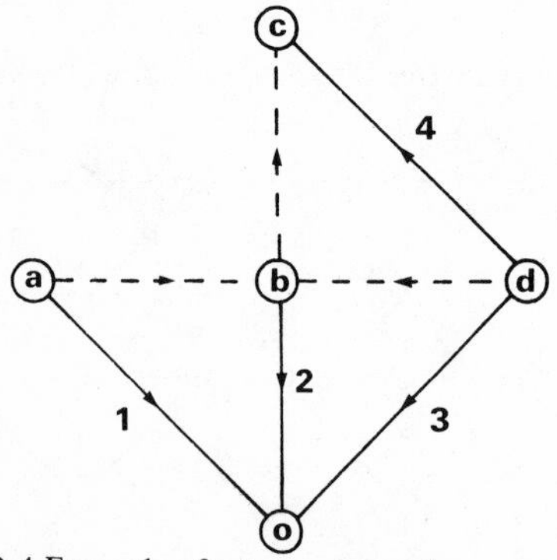

Fig. 2.4 Example of a tree sub-graph

the tree incidence matrix $\mathbf{C}_t$ and the branch-path incidence matrix $\mathbf{A}$ are:

$$\mathbf{C}_t = \begin{array}{c} \\ a \\ b \\ c \\ d \end{array} \begin{array}{c} \begin{array}{cccc} 1 & 2 & 3 & 4 \end{array} \\ \begin{bmatrix} 1 & \cdot & \cdot & \cdot \\ \cdot & 1 & \cdot & \cdot \\ \cdot & \cdot & \cdot & -1 \\ \cdot & \cdot & 1 & 1 \end{bmatrix} \end{array}, \quad \mathbf{A} = \begin{array}{c} \\ a \\ b \\ c \\ d \end{array} \begin{array}{c} \begin{array}{cccc} 1 & 2 & 3 & 4 \end{array} \\ \begin{bmatrix} 1 & \cdot & \cdot & \cdot \\ \cdot & 1 & \cdot & \cdot \\ \cdot & \cdot & 1 & -1 \\ \cdot & \cdot & 1 & \cdot \end{bmatrix} \end{array}$$

For any network, the branch-path incidence matrix is related to the tree branch-nodal incidence matrix by:

$$\mathbf{C}_t \mathbf{A}^T = \mathbf{U} \tag{2.1}$$

where $\mathbf{U}$ is a unit matrix

or $$\mathbf{A}^T = \mathbf{C}_t^{-1} \tag{2.2}$$

This relation can be very useful because it is much simpler to transpose a matrix than to invert one.

2.2.5 Branch-loop incidence matrix

Any closed path passing only once through any branch is defined as a *loop*. The number of independent loops is equal to the number of independent loop equations. A *basic loop* may consist of any number of branches of the tree but one, and only one, co-tree branch. Therefore, basic loops depend on the initial definition of the tree for any given graph. Since each basic loop corresponds to one co-tree branch, the number of basic loops and therefore the number of independent loops is equal to the number of co-tree branches. The basic loops can be defined by the *basic branch-loop* incidence matrix $\mathbf{D}$ in which the element d_{ij} in row i and column j corresponds to loop i and branch j. This element is defined as:

$$d_{ij} = \begin{cases} +1, & \text{if the direction of the basic loop } i \text{ is the same as the direction of branch } j \\ -1, & \text{if the direction of the basic loop } i \text{ is opposite to the direction of branch } j \\ 0, & \text{if the basic loop } i \text{ does not include branch } j \end{cases}$$

If the direction of the basic loops is made the same as the direction of the co-tree branches and the branch-loop incidence matrix is partitioned to separate the tree from the co-tree branches, then the following matrix is obtained:

$$
\begin{aligned}
\mathbf{D} &= \left[\mathbf{D}_t \;\vdots\; \mathbf{D}_c\right] \\
&= \left[\mathbf{D}_t \;\vdots\; \mathbf{U}\;\right]
\end{aligned}
$$

since $\mathbf{D}_c$ is a unit matrix.

In the example shown in Fig. 2.2, two basic loops p and q exist. Therefore for this example the partitioned basic branch-loop incidence matrix is:

$$
\begin{array}{cc}
 & \begin{array}{ccccc} 1 & 2 & 3 & 4 & 5 \end{array} \\
\mathbf{D} = \begin{array}{c} p \\ \\ q \end{array} & \left[\begin{array}{ccc:cc} -1 & 1 & \cdot & 1 & \cdot \\ & & & & \\ \cdot & -1 & 1 & \cdot & 1 \end{array}\right] \\
 & \begin{array}{cc} \underbrace{\hphantom{-1 \quad 1 \quad 1}}_{\text{tree}} & \underbrace{\hphantom{1 \quad 1}}_{\text{co-tree}} \end{array}
\end{array}
$$

The matrix $\mathbf{D}_t$ may be expressed in terms of the branch-nodal incidence matrices $\mathbf{C}_t$ and $\mathbf{C}_c$ and the branch-path incidence matrix $\mathbf{A}$ by:

$$
\mathbf{D}_t^T = -\mathbf{C}_t^{-1}\,\mathbf{C}_c = -\mathbf{A}^T\,\mathbf{C}_c \qquad (2.3)
$$

which, for the example shown in Fig. 2.2, gives:

$$
\begin{bmatrix} -1 & \cdot \\ 1 & -1 \\ \cdot & 1 \end{bmatrix} = - \begin{bmatrix} 1 & \cdot & \cdot \\ \cdot & 1 & 1 \\ \cdot & \cdot & -1 \end{bmatrix} \begin{bmatrix} 1 & \cdot \\ -1 & \cdot \\ \cdot & 1 \end{bmatrix}
$$

Other loops which are not basic ones may be obtained for a given graph by a linear transformation of the basic loops. For the example shown in Fig. 2.2, a new row vector is obtained by adding the rows corresponding to loops p and q. This new vector is:

$$
\begin{array}{cc}
 & \begin{array}{ccccc} 1 & 2 & 3 & 4 & 5 \end{array} \\
r & \left[\begin{array}{ccccc} -1 & \cdot & 1 & 1 & 1 \end{array}\right]
\end{array}
$$

which defines the loop shown in Fig. 2.5.

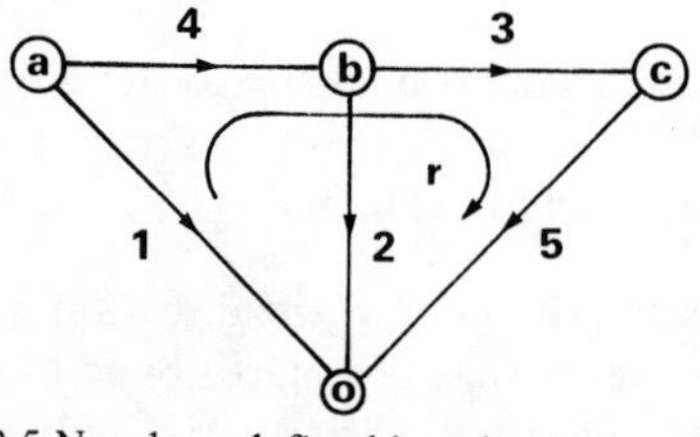

Fig. 2.5 New loop defined by a linear transformation of basic loops

In general, if one row of the branch-loop incidence matrix is replaced by a new row obtained from a linear transformation of the basic loops, a new branch-loop incidence matrix is obtained. This matrix defines new independent loops. This property of changing from one set of independent loops to another is very important when loops with special character are required.

2.3 Nodal formulation of linear network equations

2.3.1 Formulation of equations

The method described in this section illustrates the general procedure of establishing simple topological relations in matrix form and the routine steps used to assemble the functional equations of a complex system from these simple relations.

Consider the network shown in Fig. 2.6.

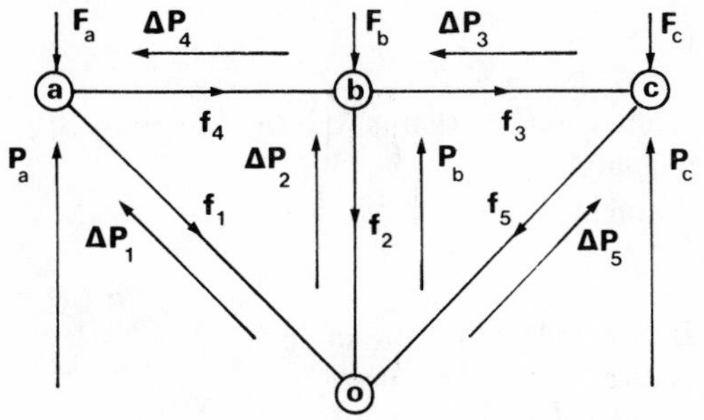

Fig. 2.6 Nodal formulation of linear network equations

There are two quantities associated with each branch: the *through* quantity f, corresponding to the flow in the branch, and the *across* quantity ΔP, corresponding to the potential difference across the branch. These may represent, for example, electric current and voltage difference or gas flow and pressure difference. For a linear network these two branch quantities are related by:

$$\Delta P = rf \tag{2.4}$$

where r is a proportionality constant of a branch and will be referred to as the branch resistance.

For a general network the number of such relations is equal to the number of branches and can be represented in matrix form by:

$$\mathbf{\Delta P} = \mathbf{Rf} \tag{2.5}$$

$$\text{where } \mathbf{R} = \begin{bmatrix} r_1 & \cdot & \cdot & \cdot \\ \cdot & r_2 & \cdot & \cdot \\ \cdot & \cdot & \ddots & \cdot \\ \cdot & \cdot & \cdot & r_b \end{bmatrix}, \quad \Delta\mathbf{P} = \begin{bmatrix} \Delta P_1 \\ \Delta P_2 \\ \vdots \\ \Delta P_b \end{bmatrix}, \quad \mathbf{f} = \begin{bmatrix} f_1 \\ f_2 \\ \vdots \\ f_b \end{bmatrix}$$

and b = total number of branches in the network.

If a nodal potential relative to a reference node is defined, for example, P_a, P_b and P_c in Fig. 2.6 with reference to node 0, then the 'across' quantity ΔP_j for any given branch j may be related to these nodal potentials by:

$$\Delta P_j = P_{j(s)} - P_{j(r)} \tag{2.6}$$

where $P_{j(s)}$ and $P_{j(r)}$ indicate the nodal potential values at the sending and receiving end respectively of branch j.

Equation 2.6 may be expressed in terms of all the defined nodal potentials by:

$$\Delta P_j = \sum_{i=1}^{n} c_{ji} P_i \qquad (j = 1, 2, \ldots, b) \tag{2.7}$$

where n = number of independent nodes, that is, the total number of nodes minus the reference node

b = total number of branches

and

$$c_{ji} = \begin{cases} +1, & \text{if branch } j \text{ is directed away from node } i \\ -1, & \text{if branch } j \text{ is directed towards node } i \\ 0, & \text{if branch } j \text{ is not connected to node } i \end{cases}$$

Equation 2.7 may be expressed in matrix form by:

$$\Delta\mathbf{P} = \mathbf{C}^{\mathrm{T}}\,\mathbf{P} \tag{2.8}$$

From the definition of the elements of $\mathbf{C}^{\mathrm{T}}$, it is evident that this matrix is the transpose of the branch-nodal incidence matrix $\mathbf{C}$, defined previously on page 6.

Kirchhoff's first law states that the algebraic sum of the flows entering and leaving any given node is equal to zero. This may be written in general form as:

$$\sum_{j=1}^{b} c_{ij} f_j = F_i \qquad (i = 1, 2, \ldots, n) \tag{2.9}$$

where n = number of independent nodes

b = total number of branches

and F_i = net flow injected at node i

Equation 2.9 can be written in matrix form as:

$$\mathbf{Cf} = \mathbf{F} \tag{2.10}$$

where **C** is the branch-nodal incidence matrix defined on page 7.

For the example shown in Fig. 2.6, the potential difference or 'across' quantity of individual branches may be expressed in terms of the nodal potentials, using equation 2.7, as:

$$\Delta P_1 = P_a$$

$$\Delta P_2 = P_b$$

$$\Delta P_3 = P_b - P_c$$

$$\Delta P_4 = P_a - P_b$$

$$\Delta P_5 = P_c$$

Using equation 2.8, these relations may be written in matrix form as:

$$\begin{bmatrix} \Delta P_1 \\ \Delta P_2 \\ \Delta P_3 \\ \Delta P_4 \\ \Delta P_5 \end{bmatrix} = \begin{bmatrix} 1 & \cdot & \cdot \\ \cdot & 1 & \cdot \\ \cdot & 1 & -1 \\ 1 & -1 & \cdot \\ \cdot & \cdot & 1 \end{bmatrix} \begin{bmatrix} P_a \\ P_b \\ P_c \end{bmatrix}$$

Similarly, using equation 2.9, the equations for nodal flows can be expressed as:

$$f_1 + f_4 = F_a$$

$$f_2 + f_3 - f_4 = F_b$$

$$-f_3 + f_5 = F_c$$

and using equation 2.10, these relations may be written in matrix form as:

$$\begin{bmatrix} 1 & \cdot & \cdot & 1 & \cdot \\ \cdot & 1 & 1 & -1 & \cdot \\ \cdot & \cdot & -1 & \cdot & 1 \end{bmatrix} \begin{bmatrix} f_1 \\ f_2 \\ f_3 \\ f_4 \\ f_5 \end{bmatrix} = \begin{bmatrix} F_a \\ F_b \\ F_c \end{bmatrix}$$

It can be seen that the coefficient matrix of this equation is identical to the branch-nodal incidence matrix of the network as shown on page 7.

Substituting f from equation 2.5 into equation 2.10 gives:

$$\mathbf{C}\,\mathbf{R}^{-1}\,\Delta\mathbf{P} = \mathbf{F} \tag{2.11}$$

and substituting $\Delta\mathbf{P}$ from equation 2.8 into equation 2.11 gives:

$$\mathbf{C}\,\mathbf{R}^{-1}\,\mathbf{C}^{\mathrm{T}}\mathbf{P} = \mathbf{F}$$

or

$$\mathbf{Y}\mathbf{P} = \mathbf{F} \tag{2.12}$$

where

$$\mathbf{Y} = \mathbf{C}\mathbf{R}^{-1}\,\mathbf{C}^{\mathrm{T}} \tag{2.13}$$

The matrix **Y** is generally known as the *nodal coefficient* matrix, and in electrical network problems is more often referred to as the *nodal admittance* matrix. For the example shown in Fig. 2.6, it is:

$$\mathbf{Y} = \begin{bmatrix} \left(\frac{1}{r_1}+\frac{1}{r_4}\right) & -\left(\frac{1}{r_4}\right) & \cdot \\ -\left(\frac{1}{r_4}\right) & \left(\frac{1}{r_2}+\frac{1}{r_3}+\frac{1}{r_4}\right) & -\left(\frac{1}{r_3}\right) \\ \cdot & -\left(\frac{1}{r_3}\right) & \left(\frac{1}{r_3}+\frac{1}{r_5}\right) \end{bmatrix}$$

2.3.2 Properties of the nodal coefficient matrix

(i) From equation 2.13,

$$\begin{aligned} \mathbf{Y}^{\mathrm{T}} &= (\mathbf{C}\mathbf{R}^{-1}\,\mathbf{C}^{\mathrm{T}})^{\mathrm{T}} \\ &= (\mathbf{C}^{\mathrm{T}})^{\mathrm{T}}\,(\mathbf{R}^{-1})^{\mathrm{T}}\,(\mathbf{C})^{\mathrm{T}} \\ &= \mathbf{C}(\mathbf{R}^{-1})^{\mathrm{T}}\,\mathbf{C}^{\mathrm{T}} \end{aligned} \tag{2.14}$$

Since **R** and therefore $\mathbf{R}^{-1}$ are diagonal matrices,

$$(\mathbf{R}^{-1})^{\mathrm{T}} = \mathbf{R}^{-1} \tag{2.15}$$

therefore, from equations 2.14 and 2.15,

$$\begin{aligned} \mathbf{Y}^{\mathrm{T}} &= \mathbf{C}\mathbf{R}^{-1}\,\mathbf{C}^{\mathrm{T}} \\ &= \mathbf{Y} \end{aligned} \tag{2.16}$$

Thus the coefficient matrix **Y** of the nodal equations is symmetrical.

(ii) The off-diagonal elements y_{ij} of **Y** can be determined from

$$y_{ij} = \mathbf{C}_{\mathrm{i}}\,\mathbf{R}^{-1}\,\mathbf{C}_{\mathrm{j}}^{\mathrm{T}} \tag{2.17}$$

where $\mathbf{C}_{\mathrm{i}}$ and $\mathbf{C}_{\mathrm{j}}$ are rows i and j of **C** respectively.

Expanding equation 2.17 gives:

$$y_{ij} = \sum_{k=1}^{b} c_{ik}\,r_k^{-1}\,c_{jk} \qquad (i, j = 1, 2, \ldots, n) \tag{2.18}$$

where b = number of branches
n = number of independent nodes

and c_{ik} and c_{jk} are the elements in row i and j respectively of the branch-nodal incidence matrix **C**.

Both c_{ik} and c_{jk} are non-zero only if branch k is connected to both nodes i and j. If branch k is connected to nodes i and j, one of the nodes is the sending end and the other is the receiving end. Therefore,

$$c_{ik} = -c_{jk}$$

$$\text{and} \quad c_{ik}\, c_{jk} = -1 \tag{2.19}$$

If branch k is not connected to both nodes i and j, c_{ik} and/or c_{jk} are zero. Therefore,

$$c_{ik}\, c_{jk} = 0 \tag{2.20}$$

Hence, from equations 2.18–2.20, the element y_{ij} is obtained by summing algebraically the reciprocal of the resistances of the branches between nodes i and j and changing its sign.

(iii) The diagonal elements y_{ii} of **Y** can be determined from:

$$y_{ii} = \mathbf{C}_i \mathbf{R}^{-1} \mathbf{C}_i^{\mathrm{T}} \tag{2.21}$$

Expanding equation 2.21 gives:

$$y_{ii} = \sum_{k=1}^{b} c_{ik}\, r_k^{-1}\, c_{ik}$$

$$= \sum_{k=1}^{b} (c_{ik})^2\, r_k^{-1} \tag{2.22}$$

For a branch k connected to node i

$$(c_{ik})^2 = 1 \tag{2.23}$$

and for a branch k not connected to node i

$$(c_{ik})^2 = 0 \tag{2.24}$$

Therefore, from equations 2.22–2.24, the diagonal element y_{ii} of **Y** is obtained by summing algebraically the reciprocal of the resistances of all branches connected to node i.

(iv) The total number of non-zero elements in the coefficient matrix **Y** is $n + 2b'$ where n is the number of independent nodes and b' is the number of branches connected between the independent nodes. Parallel branches connected between the same nodes are counted as one branch. In the example shown in Fig. 2.6, $n = 3$ and $b' = 2$.

2.4 Loop formulation of linear network equations

2.4.1 Formulation of equations

It was shown in section 2.3 that linear network equations can be formulated from a knowledge of the topology of the network and incidence matrices, using nodal injected flows. A similar set of equations can be formulated for the same network topology using the loop method. This requires that the problem is represented by potential sources in series with the branches. To illustrate this method of formulation, consider the network shown in Fig. 2.7.

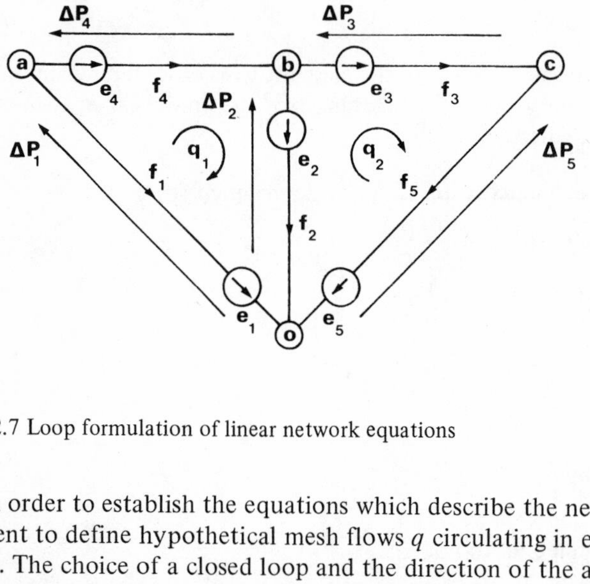

Fig. 2.7 Loop formulation of linear network equations

In order to establish the equations which describe the network it is convenient to define hypothetical mesh flows q circulating in each independent loop. The choice of a closed loop and the direction of the assumed loop flow is quite arbitrary. The number of independent loops is related to the number of nodes and branches and, as shown on page 9, is equal to the number of co-tree branches in a network.

Using the loops defined in Fig. 2.7, the flows in the individual branches can be expressed in terms of the loop flows by:

$$f_1 = -q_1$$

$$f_2 = q_1 - q_2$$

$$f_3 = q_2$$

$$f_4 = q_1$$

$$f_5 = q_2$$

These relations can be written in matrix form as:

$$\begin{bmatrix} f_1 \\ f_2 \\ f_3 \\ f_4 \\ f_5 \end{bmatrix} = \begin{bmatrix} -1 & \cdot \\ 1 & -1 \\ \cdot & 1 \\ 1 & \cdot \\ \cdot & 1 \end{bmatrix} \begin{bmatrix} q_1 \\ q_2 \end{bmatrix}$$

i.e. $\mathbf{f} = \mathbf{D}^{\mathrm{T}}\,\mathbf{Q}$ (2.25)

where **D** is the branch-loop incidence matrix previously defined on page 9.

Similarly, the equations for potential difference or 'across' quantities in any closed loop can be obtained using Kirchhoff's second law, and may be written as:

$$(r_1 f_1 - e_1) - (r_2 f_2 - e_2) - (r_4 f_4 - e_4) = 0$$

$$(r_2 f_2 - e_2) - (r_3 f_3 - e_3) - (r_5 f_5 - e_5) = 0$$

where r is the resistance of each branch.

In matrix form, these relations become:

$$\begin{bmatrix} 1 & -1 & \cdot & -1 & \cdot \\ \cdot & 1 & -1 & \cdot & -1 \end{bmatrix} \begin{bmatrix} r_1 f_1 \\ r_2 f_2 \\ r_3 f_3 \\ r_4 f_4 \\ r_5 f_5 \end{bmatrix} = \begin{bmatrix} 1 & -1 & \cdot & -1 & \cdot \\ \cdot & 1 & -1 & \cdot & -1 \end{bmatrix} \begin{bmatrix} e_1 \\ e_2 \\ e_3 \\ e_4 \\ e_5 \end{bmatrix}$$

i.e. $\mathbf{DRf} = \mathbf{De}$ (2.26)

Substituting for **f** from equation 2.25 into equation 2.26 gives:

$$\mathbf{DRD}^{\mathrm{T}}\,\mathbf{Q} = \mathbf{De} \qquad (2.27)$$

Equation 2.27 can be expressed as:

$$\mathbf{Z}\mathbf{Q} = \mathbf{E} \qquad (2.28)$$

where $\mathbf{Z} = \mathbf{DRD}^{\mathrm{T}}$ and $\mathbf{E} = \mathbf{De}$ (2.29)

The matrix **Z** is generally known as the *loop coefficient* matrix and in electrical network solutions is more often referred to as the *loop impedance* matrix. For the example shown in Fig. 2.7, it is:

$$\mathbf{Z} = \begin{bmatrix} (r_1 + r_2 + r_4) & (-r_2) \\ (-r_2) & (r_2 + r_3 + r_5) \end{bmatrix}$$

2.4.2 Properties of the loop coefficient matrix

(i) From equation 2.29,

$$\mathbf{Z}^T = (\mathbf{D}\mathbf{R}\mathbf{D}^T)^T$$

$$= \mathbf{D}\mathbf{R}^T \mathbf{D}^T \qquad (2.30)$$

Since $\mathbf{R}$ is a diagonal matrix,

$$\mathbf{R}^T = \mathbf{R} \qquad (2.31)$$

Therefore, from equations 2.30 and 2.31,

$$\mathbf{Z}^T = \mathbf{D}\mathbf{R}\mathbf{D}^T$$

$$= \mathbf{Z} \qquad (2.32)$$

Thus the coefficient matrix $\mathbf{Z}$ of the loop equations is symmetrical.
(ii) The off-diagonal elements z_{ij} of $\mathbf{Z}$ can be determined from

$$z_{ij} = \mathbf{D}_i \mathbf{R} \mathbf{D}_j^T \qquad (2.33)$$

where $\mathbf{D}_i$ and $\mathbf{D}_j$ are rows i and j of $\mathbf{D}$ respectively.
Expanding equation 2.33 gives:

$$z_{ij} = \sum_{k=1}^{b} d_{ik}\, r_k\, d_{jk} \qquad (i, j = 1, 2, \ldots, l) \qquad (2.34)$$

where b = number of branches
l = number of independent loops

and d_{ik} and d_{jk} are the elements in row i and j respectively of the branch-loop incidence matrix $\mathbf{D}$.

Both d_{ik} and d_{jk} are non-zero for the branches which are common to both loops i and j. Therefore, the off-diagonal element z_{ij} is equal to the summation of the resistances of the branches common to loops i and j if the direction of the loops is the same. If the loops through a branch are in opposite directions, the branch resistance must be multiplied by (-1) before summing.
(iii) The diagonal elements z_{ii} of $\mathbf{Z}$ can be determined from

$$z_{ii} = \mathbf{D}_i \mathbf{R} \mathbf{D}_i^T \qquad (2.35)$$

Expanding equation 2.35 gives:

$$z_{ii} = \sum_{k=1}^{b} d_{ik}\, r_k\, d_{ik}$$

$$= \sum_{k=1}^{b} (d_{ik})^2 \, r_k \tag{2.36}$$

Therefore, the diagonal element z_{ii} of $\mathbf{Z}$ is obtained by summing algebraically the resistances of all branches in loop i.

(iv) The total number of non-zero elements in the coefficient matrix $\mathbf{Z}$ depends on the initial definition of the loops and is equal to the number of loops plus twice the number of incident loop pairs. In the example shown in Fig. 2.7, the number of loops is two and the number of incident loop pairs is one. Therefore the number of non-zero elements is four.

2.5 Transformation of nodal flows to loop potential sources

The coefficient matrices can be established by inspection using the rules outlined in sections 2.3 and 2.4. In general this is the quickest and most convenient way of formulating the nodal or loop equations. Difficulties are sometimes encountered in the solution of some problems, however, which can be overcome by formulating the equations using one method, and, before or during the solution, transforming the representation of the problem from this formulation to the other. It was seen in section 2.1, for example, that transformation can be useful for the solution of non-linear equations.

The use of incidence matrices in such transformations can be of considerable help. To illustrate one type of transformation, consider the nodal network shown in Fig. 2.8

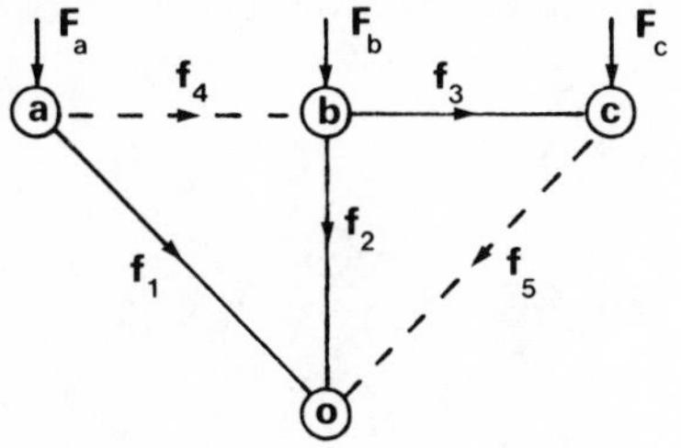

Fig. 2.8 Network with injected nodal flows

For such a network, the relation between nodal flows and branch flows was given by equation 2.10 as:

$$\mathbf{Cf} = \mathbf{F} \tag{2.10}$$

where $\mathbf{C}$ = branch-nodal incidence matrix
$\mathbf{f}$ = column vector of branch flows
$\mathbf{F}$ = column vector of nodal flows

Equation 2.10 can be partitioned in terms of flows in the tree branches $\mathbf{f}_t$ and co-tree branches $\mathbf{f}_c$ as follows:

$$\left[\mathbf{C}_t \;\vdots\; \mathbf{C}_c \right] \begin{bmatrix} \mathbf{f}_t \\ \hdashline \mathbf{f}_c \end{bmatrix} = \mathbf{F} \tag{2.37}$$

Expanding equation 2.37 gives:

$$\mathbf{C}_t \mathbf{f}_t + \mathbf{C}_c \mathbf{f}_c = \mathbf{F} \tag{2.38}$$

Since $\mathbf{C}_t$ is a square non-singular matrix (page 8), equation 2.38 can be written:

$$\mathbf{f}_t = \mathbf{C}_t^{-1}\, \mathbf{F} - \mathbf{C}_t^{-1}\, \mathbf{C}_c \mathbf{f}_c \tag{2.39}$$

The flows **f** in the individual branches are related to the potential differences $\Delta\mathbf{P}$ across the branches by equation 2.5:

$$\Delta\mathbf{P} = \mathbf{R}\mathbf{f} \qquad [2.5]$$

where **R** = diagonal matrix of individual branch resistances.

Partitioning equation 2.5 into the tree and co-tree branches gives:

$$\Delta\mathbf{P}_t = \mathbf{R}_t\, \mathbf{f}_t \tag{2.40}$$

and $\Delta\mathbf{P}_c = \mathbf{R}_c\, \mathbf{f}_c$ (2.41)

Substituting for $\mathbf{f}_t$ from equation 2.39 into 2.40 gives:

$$\Delta\mathbf{P}_t = \mathbf{R}_t\, \mathbf{C}_t^{-1}\, \mathbf{F} - \mathbf{R}_t\, \mathbf{C}_t^{-1}\, \mathbf{C}_c \mathbf{f}_c \tag{2.42}$$

Equations 2.41 and 2.42 are the fundamental equations for transforming from a nodal formulation to a loop formulation. It is desirable, however, to present an interpretation of these equations and to indicate how they relate to the physical behaviour of the network. To do this, consider the network shown in Fig. 2.9.

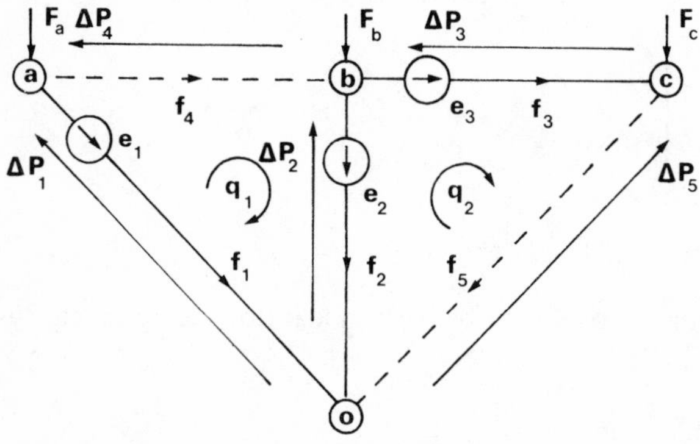

Fig. 2.9 Transformation between nodal and loop formulation

From Fig. 2.9 it is clear that the mesh flows q in the basic loops only are equal to the flows in the co-tree branches f_c,

i.e. $\mathbf{Q} = \mathbf{f}_c$ (2.43)

Therefore, equation 2.41 represents not only the potential difference across each co-tree branch using the nodal formulation but also the potential difference using the basic-loop formulation, and a direct equivalence exists;

i.e. $\Delta P_c = R_c f_c = R_c Q$ (2.44)

Also, equation 2.42 can be written using equations 2.3 and 2.43 as:

$$\Delta P_t = R_t C_t^{-1} F + R_t D_t^T Q \tag{2.45}$$

In equation 2.45, the second term on the right-hand side represents the potential drop across the tree branches due to the mesh flows in the basic loops, since from equation 2.25:

$$f_t = D_t^T Q \tag{2.46}$$

Therefore, since the total potential difference is given by ΔP_t, the first term on the right-hand side of equation 2.45 represents a potential source in series with the tree branches, and equation 2.45 may be written as:

$$\Delta P_t = - e_t + R_t D_t^T Q \tag{2.47}$$

where $e_t = - R_t C_t^{-1} F$ (2.48)

The negative sign in equation 2.48 appears due to the original definition of the elements of the matrix C.

Equation 2.42, therefore, represents not only the potential difference across each tree branch using the nodal formulation but also the potential difference using the basic-loop formulation, and again a direct equivalence exists;

i.e. $\Delta P_t = R_t C_t^{-1} F - R_t C_t^{-1} C_c f_c = - e_t + R_t D_t^T Q$ (2.49)

By reversing the process it can be shown that all potential sources in series with the branches can be transformed to injected nodal flows. This reverse process is equivalent to Norton's theorem for electrical networks.

2.6 Sparsity of the network equations

2.6.1 Sparsity coefficient

Most physical problems are characterized by the fact that they are not completely interdependent, that is, any one element is not directly linked to all the other elements. In mechanical structures or in a flow network, for example, there may be only two or three branches connected to each node. Therefore the matrix equations describing such problems contain a large proportion of zero elements and are known as sparse equations. The effect of sparsity is very important and can be very beneficial in the computational efficiency of the network solution. The *sparsity coefficient* of a matrix is defined as the ratio between the number of zero elements and the total number of elements in the matrix.

2.6.2 Sparsity of the nodal coefficient matrix

Let n be the number of independent nodes and b' the number of branches connected between the independent nodes, that is, excluding those connected to the reference node. Then the total number of non-zero elements in the nodal coefficient matrix **Y** is:

$$n + 2b'$$

The total number of elements in the matrix **Y** is:

$$n^2$$

Hence the sparsity coefficient (s.c.) of the nodal coefficient matrix **Y** is:

$$\text{s.c.} = \frac{n^2 - (n + 2b')}{n^2}$$

$$= 1 - \frac{n + 2b'}{n^2} \tag{2.50}$$

As an example, consider a practical network with 1000 nodes and 1500 branches. The sparsity coefficient of this network is:

$$\text{s.c.} = 1 - \frac{1000 + 2 \times 1500}{1000^2}$$

$$= 0{\cdot}996$$

For most practical networks, the number of branches is approximately proportional to the number of nodes. In low-pressure gas-distribution networks, for example, the ratio between the number of branches and the number of nodes is about 1·3, and in electrical transmission systems the ratio is about 1·5. Therefore, although the total number of elements in the coefficient matrix increases quadratically with the number of nodes, the number of non-zero elements only increases linearly. Thus, if:

$$b' = \alpha n$$

where α is the proportionality factor, then:

$$\text{s.c.} = 1 - \frac{1 + 2\alpha}{n} \tag{2.51}$$

and, for any given factor α, the sparsity coefficient approaches unity as the number of nodes is increased.

An important property of the nodal coefficient matrix **Y** is that, for any given network, the sparsity coefficient depends only on the graph of the network, that is, the number of branches and number of nodes, and therefore is a constant.

2.6.3 Sparsity of the loop coefficient matrix

The number of non-zero elements in the loop coefficient matrix $\mathbf{Z}$ is:

$$b - n + 2p$$

where b = total number of branches
n = number of independent nodes, that is, excluding the reference node
p = number of incident loop pairs
since $b - n$ is the number of independent loops.

The total number of elements in the matrix $\mathbf{Z}$ is:

$$(b - n)^2$$

Hence the sparsity coefficient of the loop coefficient matrix $\mathbf{Z}$ is:

$$\text{s.c.} = \frac{(b-n)^2 - (b-n+2p)}{(b-n)^2}$$

$$= 1 - \frac{b-n+2p}{(b-n)^2} \qquad (2.52)$$

For any given network, the sparsity coefficient of the loop coefficient matrix will depend on the initial definition of the loops. Consider the network shown in Fig. 2.10(a), for example, and two alternative ways of defining the loops as shown in Figs. 2.10(b) and (c).

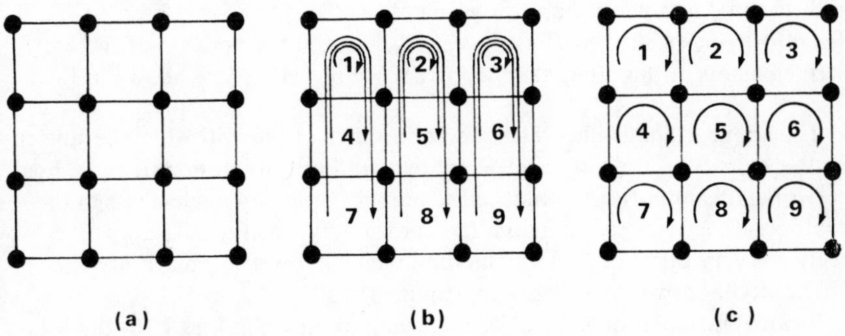

Fig. 2.10 Alternative ways of defining loops

Assuming that one of the nodes in Fig. 2.10 is the reference node, then for the loops shown in Fig. 2.10(b):

$$n = 15, \quad b = 24 \quad \text{and} \quad p = 27$$

and the total number of non-zero elements is 63.
Therefore the sparsity coefficient, s.c. = 0·222 and the position of the non-zero elements in the loop coefficient matrix is as shown in Fig. 2.11.

	1	2	3	4	5	6	7	8	9
1	X	X	·	X	X	·	X	X	·
2	X	X	X	X	X	X	X	X	X
3	·	X	X	·	X	X	·	X	X
4	X	X	·	X	X	·	X	X	·
5	X	X	X	X	X	X	X	X	X
6	·	X	X	·	X	X	·	X	X
7	X	X	·	X	X	·	X	X	·
8	X	X	X	X	X	X	X	X	X
9	·	X	X	·	X	X	·	X	X

Fig. 2.11 Position of non-zero elements in $\mathbf{Z}$ for loops defined as in Fig. 2.10 (b)

For the alternative loops as shown in Fig. 2.10(c):

$$n = 15, \quad b = 24 \quad \text{and} \quad p = 12$$

and the total number of non-zero elements is 33.
Therefore the sparsity coefficient, s.c. = 0·593 and the position of the non-zero elements in this alternative loop coefficient matrix is as shown in Fig. 2.12.

The above example illustrates the importance of the initial loop definitions on the sparsity of the loop coefficient matrix. For the solution of very large networks it is important to reduce the number of non-zero elements in the coefficient matrix to a minimum. Except for very small regular networks, however, it is very difficult to find a method which defines automatically and efficiently loops which will give maximum sparsity.

If nodal formulation is used for the example shown in Fig. 2.10, the total number of non-zero elements in the nodal coefficient matrix is 59 and the sparsity coefficient is:

s.c. = 0·693

By carefully defining the loops of the network shown in Fig. 2.10, the number of non-zero elements in the loop coefficient matrix is less than the number in the nodal coefficient matrix. With a poor choice of loops, however, the number of non-zero elements in the loop coefficient matrix can be greater. Also, although the number of equations in the nodal formulation is larger than in the loop formulation, its sparsity coefficient is smaller. This

	1	2	3	4	5	6	7	8	9
1	X	X	·	X	·	·	·	·	·
2	X	X	X	·	X	·	·	·	·
3	·	X	X	·	·	X	·	·	·
4	X	·	·	X	X	·	X	·	·
5	·	X	·	X	X	X	·	X	·
6	·	·	X	·	X	X	·	·	X
7	·	·	·	X	·	·	X	X	·
8	·	·	·	·	X	·	X	X	X
9	·	·	·	·	·	X	·	X	X

Fig. 2.12 Position of non-zero elements in $\mathbf{Z}$ for loops defined as in Fig. 2.10 (*c*)

sparsity coefficient has a significant effect on the computational efficiency of the network solution. Although the number of equations in the nodal formulation may be greater than in the loop formulation, this is usually offset due to the difficulty experienced in defining suitable loops to give maximum sparsity.

2.7 Coefficient matrices with incidence symmetry

A square matrix which is not symmetrical can be defined as *incidence symmetrical* when all the non-zero elements are replaced by unity and the resulting matrix is symmetrical. All the linear networks which have been described in this chapter have had symmetrical coefficient matrices for both the nodal and loop formulations. For some networks, however, the coefficient matrix is not necessarily symmetrical. This non-symmetry will occur, for example, in a network having flow attenuation, that is, when the flow entering one end of a branch is not equal to the flow leaving the other end of the same branch. The coefficient matrices describing such networks, however, still possess incidence symmetry and the sparsity of the networks is unaltered. For example, if two networks have the same graph but one possesses attenuation, then for every non-zero element in the coefficient matrix of one network, there is a corresponding non-zero element in the coefficient matrix of the other network. The applications of sparsity to network solutions, as discussed in the later chapters of this book, are therefore equally applicable to networks with incidence symmetry, provided the appropriate changes to the coefficient matrices are made.

With certain networks the coefficient matrix is neither symmetrical nor incidence symmetrical but has an empty position where otherwise a non-zero element could be expected. There are special techniques which can be applied to the solution of such networks, but these are outside the scope of the book. As with incidence symmetrical matrices, however, the sparsity techniques to be discussed can still be applied, if zero is inserted in the empty position and normal arithmetic operations are applied.

Bibliography

Allwood, R. J.: Matrix methods of structural analysis. In *Large Sparse Sets of Linear Equations,* pp. 17–24. Academic Press, 1971.

Balabanian, N. & Bickart, T. A.: *Electrical network theory.* John Wiley, 1969.

Baumann, R.: Sparseness in power system equations. In *Large Sparse Sets of Linear Equations,* pp. 105–25. Academic Press, 1971.

Bellman, R., Cooke, K. L. & Lockett, J. A.: *Algorithms, Graphs and Computers.* Academic Press, 1970.

Berge, C.: *The Theory of Graphs and its Applications.* John Wiley, 1964.

Brameller, A., John, M. N. & Scott, M. R.: *Practical Diakoptics for Electrical Networks.* Chapman & Hall, 1969.

Brameller, A. & Hamam, Y. M.,: Hybrid methods for the solution of piping networks. *Proc. Instn elect. Engrs* **118**, 1607–12, 1971.

Busacker, R. G. & Saaty, T. L.: *Finite Graphs and Networks.* McGraw-Hill, 1965.

Chan, S. P.: *Introductory Topological Analysis of Electrical Networks.* Holt, Rinehart & Winston, 1969.

Dulmage, A. L. & Mendelsohn, N. S.: Graphs and matrices. In *Graph Theory and Theoretical Physics,* pp. 167–277. Academic Press, 1967.

Fulkerson, D. R. & Gross, D. A.: Incidence matrices and interval graphs. *Pacif. J. Math.* **15**, 835–55, 1965.

Harary, F.: Graphs and matrices. *Symp. appl. Math. Rev.* **9**, 83–90, 1967.

Harary, F.: *Graph Theory.* Addison-Wesley, 1969.

Harary, F.: Sparse matrices and graph theory. In *Large Sparse Sets of Linear Equations,* pp. 139--50. Academic Press, 1971.

Henley, E. J. & Williams, R. A.: *Graph Theory in Modern Engineering.* Academic Press, 1973.

Hildebrand, F. B.: *Introduction to Numerical Analysis.* McGraw-Hill, 1956.

Stagg, G. W. & El-Abiad, A. H.: *Computer Methods in Power System Analysis.* McGraw-Hill, 1968.

3 Solution of Simultaneous Linear Equations

3.1 Introduction

There are many methods which can be used for solving a set of simultaneous linear equations. It is not the purpose of this book to describe such methods exhaustively, and for this the reader is referred to one of the many books on numerical methods. In order to show in later chapters how sparsity can be exploited in the solution of linear network problems, however, it is useful to indicate the various basic methods that can be used. In general, the methods can be divided into two main categories: direct and iterative.

The direct methods are based on straightforward manipulation of the equations. The solution is given within the accuracy of working and is obtained in a known and finite number of arithmetic steps. In practice, however, for large problems the methods usually require large computer storage space and long computation times, unless special techniques such as sparsity are employed.

The iterative methods are based on solving the equations by successive approximations until the results are within an acceptable accuracy limit. The problem of convergence is therefore inherent in these methods, which could suggest that direct methods should be preferred. However, iterative methods require only the coefficients of the original equations. Therefore, for matrices containing a large proportion of zero elements, such as occur in the solution of network problems, these methods are usually found to require less storage and computation time. By exploiting sparsity techniques in the direct methods, storage requirements can be significantly reduced and the computation times made shorter than those for the iterative method.

In this chapter some of the methods which are particularly applicable to the solution of general problems are examined and compared. For the solution of large sparse networks, which is the prime purpose of this book, however, special considerations should be given in order to minimize the computer storage space required and the number of arithmetical operations. These special considerations are discussed in the following chapters.

3.2 Direct methods

3.2.1 Gauss elimination

The Gauss elimination method is basically an extension of the technique generally learned at school and used for solving a set of simultaneous linear equations by hand. The technique involves eliminating one variable at a time until only one equation in one unknown remains, solving for this unknown variable and then back-substituting to solve for all the remaining variables. The Gauss method is, however, systematized and forms the basis of most of the direct methods for solving simultaneous linear equations. To illustrate the method consider the following four equations:

$$a_{11}x_1 + a_{12}x_2 + a_{13}x_3 + a_{14}x_4 = b_1 \quad (3.1a)$$

$$a_{21}x_1 + a_{22}x_2 + a_{23}x_3 + a_{24}x_4 = b_2 \quad (3.1b)$$

$$a_{31}x_1 + a_{32}x_2 + a_{33}x_3 + a_{34}x_4 = b_3 \quad (3.1c)$$

$$a_{41}x_1 + a_{42}x_2 + a_{43}x_3 + a_{44}x_4 = b_4 \quad (3.1d)$$

which may be written in matrix form as:

$$\begin{bmatrix} a_{11} & a_{12} & a_{13} & a_{14} \\ a_{21} & a_{22} & a_{23} & a_{24} \\ a_{31} & a_{32} & a_{33} & a_{34} \\ a_{41} & a_{42} & a_{43} & a_{44} \end{bmatrix} \begin{bmatrix} x_1 \\ x_2 \\ x_3 \\ x_4 \end{bmatrix} = \begin{bmatrix} b_1 \\ b_2 \\ b_3 \\ b_4 \end{bmatrix}$$

The elimination process is commenced by choosing one equation, defined as the *pivotal equation*, and, from this equation, choosing the variable to be eliminated. The coefficient of this variable is termed the *pivot*. The best coefficient to choose as a pivot is the largest one in the equation, because this tends to lead to better numerical accuracy with digital computers. In the method described below, however, the diagonal element in the pivotal equation being considered is chosen as the pivot because this leads to a neater and more systematic process of elimination. Also, for the majority of engineering problems, this choice is sufficiently accurate.

To eliminate variable x_1 from equations 3.1b, c and d,

(i) multiply equation 3.1a by $\dfrac{a_{21}}{a_{11}}$ and subtract from equation 3.1b,

(ii) multiply equation 3.1a by $\dfrac{a_{31}}{a_{11}}$ and subtract from equation 3.1c,

(iii) multiply equation 3.1a by $\dfrac{a_{41}}{a_{11}}$ and subtract from equation 3.1d.

Therefore, equations 3.1 become:

$$a_{11}x_1 + a_{12}x_2 + a_{13}x_3 + a_{14}x_4 = b_1 \tag{3.2a}$$

$$0 + a_{22}^{(1)}x_2 + a_{23}^{(1)}x_3 + a_{24}^{(1)}x_4 = b_2^{(1)} \tag{3.2b}$$

$$0 + a_{32}^{(1)}x_2 + a_{33}^{(1)}x_3 + a_{34}^{(1)}x_4 = b_3^{(1)} \tag{3.2c}$$

$$0 + a_{42}^{(1)}x_2 + a_{43}^{(1)}x_3 + a_{44}^{(1)}x_4 = b_4^{(1)} \tag{3.2d}$$

where:

$$a_{22}^{(1)} = a_{22} - \frac{a_{21}a_{12}}{a_{11}},\ a_{23}^{(1)} = a_{23} - \frac{a_{21}a_{13}}{a_{11}},$$

$$a_{24}^{(1)} = a_{24} - \frac{a_{21}a_{14}}{a_{11}},\ b_2^{(1)} = b_2 - \frac{a_{21}b_1}{a_{11}}$$

$$a_{32}^{(1)} = a_{32} - \frac{a_{31}a_{12}}{a_{11}},\ a_{33}^{(1)} = a_{33} - \frac{a_{31}a_{13}}{a_{11}},$$

$$a_{34}^{(1)} = a_{34} - \frac{a_{31}a_{14}}{a_{11}},\ b_3^{(1)} = b_3 - \frac{a_{31}b_1}{a_{11}}$$

$$a_{42}^{(1)} = a_{42} - \frac{a_{41}a_{12}}{a_{11}},\ a_{43}^{(1)} = a_{43} - \frac{a_{41}a_{13}}{a_{11}},$$

$$a_{44}^{(1)} = a_{44} - \frac{a_{41}a_{14}}{a_{11}},\ b_4^{(1)} = b_4 - \frac{a_{41}b_1}{a_{11}}$$

This process of elimination reduces all the coefficients of the variable x_1 below the pivot a_{11} to zero, without losing any of the information about the inter-relation of the variables.

To eliminate the variable x_2 from equations 3.2c and d, the above procedure is repeated, that is,

(i) multiply equation 3.2b by $\dfrac{a_{32}^{(1)}}{a_{22}^{(1)}}$ and subtract from equation 3.2c

(ii) multiply equation 3.2b by $\dfrac{a_{42}^{(1)}}{a_{22}^{(1)}}$ and subtract from equation 3.2d.

Therefore, equations 3.2 become:

$$a_{11}x_1 + a_{12}x_2 + a_{13}x_3 + a_{14}x_4 = b_1 \tag{3.3a}$$

$$0 + a_{22}^{(1)}x_2 + a_{23}^{(1)}x_3 + a_{24}^{(1)}x_4 = b_2^{(1)} \tag{3.3b}$$

$$0 + 0 + a_{33}^{(2)}x_3 + a_{34}^{(2)}x_4 = b_3^{(2)} \tag{3.3c}$$

$$0 + 0 + a_{43}^{(2)}x_3 + a_{44}^{(2)}x_4 = b_4^{(2)} \tag{3.3d}$$

where:

$$a_{33}^{(2)} = a_{33}^{(1)} - \frac{a_{32}^{(1)} a_{23}^{(1)}}{a_{22}^{(1)}}, \quad a_{34}^{(2)} = a_{34}^{(1)} - \frac{a_{32}^{(1)} a_{24}^{(1)}}{a_{22}^{(1)}},$$

$$b_3^{(2)} = b_3^{(1)} - \frac{a_{32}^{(1)} b_2^{(1)}}{a_{22}^{(1)}}$$

$$a_{43}^{(2)} = a_{43}^{(1)} - \frac{a_{42}^{(1)} a_{23}^{(1)}}{a_{22}^{(1)}}, \quad a_{44}^{(2)} = a_{44}^{(1)} - \frac{a_{42}^{(1)} a_{24}^{(1)}}{a_{22}^{(1)}},$$

$$b_4^{(2)} = b_4^{(1)} - \frac{a_{42}^{(1)} b_2^{(1)}}{a_{22}^{(1)}}$$

Similarly, by eliminating the variable x_3 from equation 3.3d, equations 3.3 become:

$$a_{11}x_1 + a_{12}x_2 + a_{13}x_3 + a_{14}x_4 = b_1 \tag{3.4a}$$

$$0 + a_{22}^{(1)}x_2 + a_{23}^{(1)}x_3 + a_{24}^{(1)}x_4 = b_2^{(1)} \tag{3.4b}$$

$$0 + 0 + a_{33}^{(2)}x_3 + a_{34}^{(2)}x_4 = b_3^{(2)} \tag{3.4c}$$

$$0 + 0 + 0 + a_{44}^{(3)}x_4 = b_4^{(3)} \tag{3.4d}$$

where:

$$a_{44}^{(3)} = a_{44}^{(2)} - \frac{a_{43}^{(2)} a_{34}^{(2)}}{a_{33}^{(2)}}, \qquad b_4^{(3)} = b_4^{(2)} - \frac{a_{43}^{(2)} b_3^{(2)}}{a_{33}^{(2)}}$$

The effect of this elimination process has been to *triangulate* the original equations (3.1) into the *upper-triangular* form. The variables x_4, x_3, x_2 and x_1 can be determined from these triangulated equations (3.4) by successive back-substitutions as follows:

$$x_4 = \frac{b_4^{(3)}}{a_{44}^{(3)}}$$

$$x_3 = \frac{b_3^{(2)}}{a_{33}^{(2)}} - \frac{a_{34}^{(2)}x_4}{a_{33}^{(2)}}$$

$$x_2 = \frac{b_2^{(1)}}{a_{22}^{(1)}} - \frac{a_{24}^{(1)}x_4}{a_{22}^{(1)}} - \frac{a_{23}^{(1)}x_3}{a_{22}^{(1)}}$$

$$x_1 = \frac{b_1}{a_{11}} - \frac{a_{14}x_4}{a_{11}} - \frac{a_{13}x_3}{a_{11}} - \frac{a_{12}x_2}{a_{11}}$$

The elimination process therefore consists of two parts: firstly, the forward elimination to give the upper-triangular form of the equations, and secondly, the back-substitution to obtain the values of the unknown variables. For a relatively large system having n equations, the reduction to the upper-triangular form takes about $\frac{n^3}{3}$ operations and the back-substitution about $\frac{n^2}{2}$ operations.

To illustrate the process described above, consider the following numerical example in matrix form:

$$\begin{bmatrix} 3 & -1 & -1 & 0 \\ -1 & 2 & 0 & 0 \\ -1 & 0 & 2 & -1 \\ 0 & 0 & -1 & 1 \end{bmatrix} \begin{bmatrix} x_1 \\ x_2 \\ x_3 \\ x_4 \end{bmatrix} = \begin{bmatrix} 1 \\ 1 \\ 1 \\ 1 \end{bmatrix}$$

Eliminating the coefficients of x_1 from rows 2–4 gives:

$$\begin{bmatrix} 3 & -1 & -1 & 0 \\ 0 & \frac{5}{3} & -\frac{1}{3} & 0 \\ 0 & -\frac{1}{3} & \frac{5}{3} & -1 \\ 0 & 0 & -1 & 1 \end{bmatrix} \begin{bmatrix} x_1 \\ x_2 \\ x_3 \\ x_4 \end{bmatrix} = \begin{bmatrix} 1 \\ \frac{4}{3} \\ \frac{4}{3} \\ 1 \end{bmatrix}$$

Eliminating the coefficients of x_2 from rows 3 and 4 gives:

$$\begin{bmatrix} 3 & -1 & -1 & 0 \\ 0 & \frac{5}{3} & -\frac{1}{3} & 0 \\ 0 & 0 & \frac{8}{5} & -1 \\ 0 & 0 & -1 & 1 \end{bmatrix} \begin{bmatrix} x_1 \\ x_2 \\ x_3 \\ x_4 \end{bmatrix} = \begin{bmatrix} 1 \\ \frac{4}{3} \\ \frac{8}{5} \\ 1 \end{bmatrix}$$

Eliminating the coefficient of x_3 from row 4 gives:

$$\begin{bmatrix} 3 & -1 & -1 & 0 \\ 0 & \frac{5}{3} & -\frac{1}{3} & 0 \\ 0 & 0 & \frac{8}{5} & -1 \\ 0 & 0 & 0 & \frac{3}{8} \end{bmatrix} \begin{bmatrix} x_1 \\ x_2 \\ x_3 \\ x_4 \end{bmatrix} = \begin{bmatrix} 1 \\ \frac{4}{3} \\ \frac{8}{5} \\ 2 \end{bmatrix}$$

From which, by back-substitution, the solution for the variables is:

$$x_4 = 2\left(\frac{8}{3}\right) \quad = \frac{16}{3}$$

$$x_3 = \frac{5}{8}\left(\frac{8}{5} + x_4\right) = \frac{13}{3}$$

$$x_2 = \frac{3}{5}\left(\frac{4}{3} + \frac{x_3}{3}\right) = \frac{5}{3}$$

$$x_1 = \frac{1}{3}(1 + x_2 + x_3) = \frac{7}{3}$$

3.2.2 Crout elimination

The Crout elimination method, like most of the other direct methods, is a simple modification of the Gauss method. The two main differences are that, in the Crout method, each pivotal equation is divided by its pivot and the coefficients of the variables are eliminated row by row instead of column by column.

To illustrate the method, consider the same set of linear equations shown in equations 3.1.

The first stage of elimination consists of:

(i) divide equation 3.1a by a_{11}

(ii) multiply the modified pivoted equations by a_{21} and subtract from equation 3.1b.

Therefore, at the end of the first stage of elimination, equations 3.1 become:

$$x_1 + a_{12}^{(1)}x_2 + a_{13}^{(1)}x_3 + a_{14}^{(1)}x_4 = b_1^{(1)} \tag{3.5a}$$

$$0 \ + a_{22}^{(1)}x_2 + a_{23}^{(1)}x_3 + a_{24}^{(1)}x_4 = b_2^{(1)} \tag{3.5b}$$

$$a_{31}x_1 + a_{32}x_2 + a_{33}x_3 + a_{34}x_4 = b_3 \tag{3.5c}$$

$$a_{41}x_1 + a_{42}x_2 + a_{43}x_3 + a_{44}x_4 = b_4 \tag{3.5d}$$

where:

$$a_{12}^{(1)} = \frac{a_{12}}{a_{11}},\ a_{13}^{(1)} = \frac{a_{13}}{a_{11}},\ a_{14}^{(1)} = \frac{a_{14}}{a_{11}},\ b_1^{(1)} = \frac{b_1}{a_{11}}$$

and $a_{22}^{(1)}$, $a_{23}^{(1)}$, $a_{24}^{(1)}$, $b_2^{(1)}$ are as shown on page 29.

The next stage of the process involves eliminating the coefficients of x_1 and x_2 in equation 3.5c. To achieve this elimination:

(i) divide equation 3.5b by $a_{22}^{(1)}$, which gives:

$$0 + x_2 + a_{23}^{(2)}x_3 + a_{24}^{(2)}x_4 = b_2^{(2)} \tag{3.6a}$$

where:

$$a_{23}^{(2)} = \frac{a_{23}^{(1)}}{a_{22}^{(1)}}, a_{24}^{(2)} = \frac{a_{24}^{(1)}}{a_{22}^{(1)}}, b_2^{(2)} = \frac{b_2^{(1)}}{a_{22}^{(1)}}$$

(ii) multiply equation 3.5a, by a_{31} and subtract from equation 3.5c, which gives:

$$0 + a_{32}^{(1)}x_2 + a_{33}^{(1)}x_3 + a_{34}^{(1)}x_4 = b_3^{(1)} \tag{3.6b}$$

where:

$$a_{32}^{(1)} = a_{32} - a_{31}a_{12}^{(1)}, a_{33}^{(1)} = a_{33} - a_{31}a_{13}^{(1)}, a_{34}^{(1)} = a_{34} - a_{31}a_{14}^{(1)},$$
$$b_3^{(1)} = b_3 - a_{31}b_1^{(1)}$$

the values of $a_{32}^{(1)}$, etc., are numerically the same as shown on page 29.
(iii) multiply equation 3.6a by $a_{32}^{(1)}$ and subtract from equation 3.6b, which gives:

$$0 + 0 + a_{33}^{(2)}x_3 + a_{34}^{(2)}x_4 = b_3^{(2)} \tag{3.6c}$$

where:

$$a_{33}^{(2)} = a_{33}^{(1)} - a_{32}^{(1)}a_{23}^{(2)}, a_{34}^{(2)} = a_{34}^{(1)} - a_{32}^{(1)}a_{24}^{(2)},$$
$$b_3^{(2)} = b_3^{(1)} - a_{32}^{(1)}b_2^{(2)}$$

Therefore, at the end of the second stage of elimination, equations 3.5 become:

$$x_1 + a_{12}^{(1)}x_2 + a_{13}^{(1)}x_3 + a_{14}^{(1)}x_4 = b_1^{(1)} \tag{3.7a}$$
$$0 + \quad x_2 \quad + a_{23}^{(2)}x_3 + a_{24}^{(2)}x_4 = b_2^{(2)} \tag{3.7b}$$
$$0 + \quad 0 \quad + a_{33}^{(2)}x_3 + a_{34}^{(2)}x_4 = b_3^{(2)} \tag{3.7c}$$
$$a_{41}x_1 + a_{42}x_2 + a_{43}x_3 + a_{44}x_4 = b_4 \tag{3.7d}$$

which can be written in matrix form as:

$$\begin{bmatrix} 1 & a_{12}^{(1)} & a_{13}^{(1)} & a_{14}^{(1)} \\ 0 & 1 & a_{23}^{(2)} & a_{24}^{(2)} \\ 0 & 0 & a_{33}^{(2)} & a_{34}^{(2)} \\ a_{41} & a_{42} & a_{43} & a_{44} \end{bmatrix} \begin{bmatrix} x_1 \\ x_2 \\ x_3 \\ x_4 \end{bmatrix} = \begin{bmatrix} b_1^{(1)} \\ b_2^{(2)} \\ b_3^{(2)} \\ b_4 \end{bmatrix}$$

From the coefficient matrix of this equation, it is clearly seen that the Crout elimination method reduces to zero the coefficients to the left of the diagonal in a row by row manner and creates a diagonal whose elements are unity. This elimination process is continued until the equations, and therefore the coefficient matrix, are reduced to the upper triangular form from which the solution of all the variables can be determined by back-substitution.

To illustrate the Crout elimination method, consider the same numerical example used before (page 31). This was:

$$\begin{bmatrix} 3 & -1 & -1 & 0 \\ -1 & 2 & 0 & 0 \\ -1 & 0 & 2 & -1 \\ 0 & 0 & -1 & 1 \end{bmatrix} \begin{bmatrix} x_1 \\ x_2 \\ x_3 \\ x_4 \end{bmatrix} = \begin{bmatrix} 1 \\ 1 \\ 1 \\ 1 \end{bmatrix}$$

Eliminating the coefficient of x_1 from row 2 gives:

$$\begin{bmatrix} 1 & -\frac{1}{3} & -\frac{1}{3} & 0 \\ -1 & 2 & 0 & 0 \\ -1 & 0 & 2 & -1 \\ 0 & 0 & -1 & 1 \end{bmatrix} \begin{bmatrix} x_1 \\ x_2 \\ x_3 \\ x_4 \end{bmatrix} = \begin{bmatrix} \frac{1}{3} \\ 1 \\ 1 \\ 1 \end{bmatrix}, \quad \begin{bmatrix} 1 & -\frac{1}{3} & -\frac{1}{3} & 0 \\ 0 & \frac{5}{3} & -\frac{1}{3} & 0 \\ -1 & 0 & 2 & -1 \\ 0 & 0 & -1 & 1 \end{bmatrix} \begin{bmatrix} x_1 \\ x_2 \\ x_3 \\ x_4 \end{bmatrix} = \begin{bmatrix} \frac{1}{3} \\ \frac{4}{3} \\ 1 \\ 1 \end{bmatrix}$$

Eliminating the coefficients of x_1 and x_2 from row 3 gives:

$$\begin{bmatrix} 1 & -\frac{1}{3} & -\frac{1}{3} & 0 \\ 0 & 1 & -\frac{1}{5} & 0 \\ -1 & 0 & 2 & -1 \\ 0 & 0 & -1 & 1 \end{bmatrix} \begin{bmatrix} x_1 \\ x_2 \\ x_3 \\ x_4 \end{bmatrix} = \begin{bmatrix} \frac{1}{3} \\ \frac{4}{5} \\ 1 \\ 1 \end{bmatrix}, \quad \begin{bmatrix} 1 & -\frac{1}{3} & -\frac{1}{3} & 0 \\ 0 & 1 & -\frac{1}{5} & 0 \\ 0 & 0 & \frac{8}{5} & -1 \\ 0 & 0 & -1 & 1 \end{bmatrix} \begin{bmatrix} x_1 \\ x_2 \\ x_3 \\ x_4 \end{bmatrix} = \begin{bmatrix} \frac{1}{3} \\ \frac{4}{5} \\ \frac{8}{5} \\ 1 \end{bmatrix}$$

Eliminating the coefficients of x_1, x_2 and x_3 from row 4 gives:

$$\begin{bmatrix} 1 & -\frac{1}{3} & -\frac{1}{3} & 0 \\ 0 & 1 & -\frac{1}{5} & 0 \\ 0 & 0 & 1 & -\frac{5}{8} \\ 0 & 0 & -1 & 1 \end{bmatrix} \begin{bmatrix} x_1 \\ x_2 \\ x_3 \\ x_4 \end{bmatrix} = \begin{bmatrix} \frac{1}{3} \\ \frac{4}{5} \\ 1 \\ 1 \end{bmatrix}, \quad \begin{bmatrix} 1 & -\frac{1}{3} & -\frac{1}{3} & 0 \\ 0 & 1 & -\frac{1}{5} & 0 \\ 0 & 0 & 1 & -\frac{5}{8} \\ 0 & 0 & 0 & \frac{3}{8} \end{bmatrix} \begin{bmatrix} x_1 \\ x_2 \\ x_3 \\ x_4 \end{bmatrix} = \begin{bmatrix} \frac{1}{3} \\ \frac{4}{5} \\ 1 \\ 2 \end{bmatrix}$$

Normalizing the last equation gives:

$$\begin{bmatrix} 1 & -\frac{1}{3} & -\frac{1}{3} & 0 \\ 0 & 1 & -\frac{1}{5} & 0 \\ 0 & 0 & 1 & -\frac{5}{8} \\ 0 & 0 & 0 & 1 \end{bmatrix} \begin{bmatrix} x_1 \\ x_2 \\ x_3 \\ x_4 \end{bmatrix} = \begin{bmatrix} \frac{1}{3} \\ \frac{4}{5} \\ 1 \\ \frac{16}{3} \end{bmatrix}$$

From which, by back-substitution, the solution for the variables is:

$$x_4 = \frac{16}{3}$$

$$x_3 = 1 + \frac{5}{8}x_4 = \frac{13}{3}$$

$$x_2 = \frac{4}{5} + \frac{1}{5}x_3 = \frac{5}{3}$$

$$x_1 = \frac{1}{3} + \frac{x_2}{3} + \frac{x_3}{3} = \frac{7}{3}$$

3.2.3 Matrix inversion

There are many technological investigations which require a number of solutions with the same or only small changes in the coefficients of the variables. This can be achieved by the repeated application of an elimination method such as described previously. However, it may be more convenient to use matrix inversion techniques. Also, there are many engineering problems which are more easily solved or handled if an inverse matrix is used in subsequent calculations.

The equation

$$\mathbf{AX} = \mathbf{b} \tag{3.8}$$

can be solved by obtaining the inverse of the matrix **A** and calculating **X** from:

$$\mathbf{X} = \mathbf{A}^{-1}\mathbf{b} \tag{3.9}$$

There are many available methods which can be used for the explicit evaluation of $\mathbf{A}^{-1}$. One of the simplest and most efficient methods for calculating the inverse of large matrices is the Gauss–Jordan or complete elimination method. This method, which is an extension of the elimination process described in the previous sections, requires no extra working storage and maintains reasonable accuracy. It is most easily understood by considering the following two equations:

$$a_{11}x_1 + a_{12}x_2 = b_1 \tag{3.10a}$$

$$a_{21}x_1 + a_{22}x_2 = b_2 \tag{3.10b}$$

which, in matrix form is:

$$\begin{bmatrix} a_{11} a_{12} \\ a_{21} a_{22} \end{bmatrix} \begin{bmatrix} x_1 \\ x_2 \end{bmatrix} = \begin{bmatrix} b_1 \\ b_2 \end{bmatrix}$$

Dividing equation 3.10b by a_{22} and re-arranging gives:

$$x_2 = a_{22}{}^{-1}(b_2 - a_{21}x_1) \tag{3.11}$$

Substituting equation 3.11 into equation 3.10a gives:

$$(a_{11} - a_{12}a_{22}{}^{-1}\, a_{21})\, x_1 + a_{12}a_{22}{}^{-1}b_2 = b_1 \tag{3.12}$$

Equations 3.12 and 3.11 may be written in matrix form as:

$$\begin{bmatrix} a_{11}{}^{(1)} a_{12}{}^{(1)} \\ a_{21}{}^{(1)} a_{22}{}^{(1)} \end{bmatrix} \begin{bmatrix} x_1 \\ b_2 \end{bmatrix} = \begin{bmatrix} b_1 \\ x_2 \end{bmatrix}$$

where:

$$a_{11}{}^{(1)} = a_{11} - a_{12}a_{22}{}^{-1}a_{21} \qquad a_{12}{}^{(1)} = a_{12}a_{22}{}^{-1}$$

$$a_{21}{}^{(1)} = -a_{22}{}^{(1)}a_{21} \qquad a_{22}{}^{(1)} = a_{22}{}^{-1}$$

The new equations are of the same form as the original equations but with x_2 and b_2 interchanged. This process can be repeated to interchange x_1 and b_1 and give:

$$\begin{bmatrix} a_{11}{}^{(2)} a_{12}{}^{(2)} \\ a_{21}{}^{(2)} a_{22}{}^{(2)} \end{bmatrix} \begin{bmatrix} b_1 \\ b_2 \end{bmatrix} = \begin{bmatrix} x_1 \\ x_2 \end{bmatrix}$$

which, in a general form, may be written as:

$$\mathbf{Cb} = \mathbf{X} \tag{3.13}$$

By comparing equations 3.9 and 3.13, it is evident that **C** is the inverse of **A**, that is:

$$\mathbf{C} = \mathbf{A}^{-1} \tag{3.14}$$

The order in which the interchange process is carried out is unimportant and an identical result would have been achieved by interchanging x_1 with b_1 before x_2 with b_2. Also, although the example shown above is trivial, the principle is valid for any number of unknown variables; the steps of the interchanging being repeated until complete inversion is achieved. This method of inversion is simplified by applying the following set of rules:

(i) $a'_{ii} = \dfrac{1}{a_{ii}}$

(ii) $a'_{ji} = a_{ji}a'_{ii}$ for all $j \neq i$

(iii) $a'_{jk} = a_{jk} - a'_{ji}\, a_{ik}$ for all $j \neq i,\ k \neq i$

(iv) $a'_{ik} = -a'_{ii}a_{ik}$ for all $k \neq i$

where a_{ii} is the diagonal element being used as the pivot and a' is the new element of the coefficient matrix after each step in the inversion process, which can be stored in the same position as the previous value.

The process is repeated for all the diagonal elements a_{ii} of the coefficient matrix, each repeated step starting from the results obtained from the previous step. Although the process can be carried out in any order, it is usually sufficient in most engineering problems to use the natural order of the equations and step $i = 1, 2, \ldots, n$. This progression leads to the following very simple FORTRAN program:

```
C   REPLACE MATRIX A OF ORDER NXN BY ITS INVERSE
      DO 6 I = 1,N
      A(I,I) = 1.0/A(I,I)
      DO 5 J = 1,N
        IF (J-I) 1, 5, 1
  1   A(J,I) = A(J,I)*A(I,I)
      DO 4 K = 1,N
        IF(K-I) 2, 4, 2
  2   A(J,K) = A(J,K) - A(J,I)*A(I,K)
        IF(J-N) 4, 3, 4
  3   A(I,K) = -A(I,I)*A(I,K)
  4   CONTINUE
  5   CONTINUE
  6   CONTINUE
      K = N - 1
      DO 7 J = 1,K
      A(N,J) = -A(N,N)*A(N,J)
  7   CONTINUE
```

To illustrate the sequential steps of the inverse process, consider the same numerical example used in the previous elimination methods (page 31):

$$\begin{bmatrix} 3 & -1 & -1 & 0 \\ -1 & 2 & 0 & 0 \\ -1 & 0 & 2 & -1 \\ 0 & 0 & -1 & 1 \end{bmatrix} \begin{bmatrix} x_1 \\ x_2 \\ x_3 \\ x_4 \end{bmatrix} = \begin{bmatrix} 1 \\ 1 \\ 1 \\ 1 \end{bmatrix}$$

Using the fourth diagonal element as pivot gives:

$$\begin{bmatrix} 3 & -1 & -1 & 0 \\ -1 & 2 & 0 & 0 \\ -1 & 0 & 1 & -1 \\ 0 & 0 & 1 & 1 \end{bmatrix} \begin{bmatrix} x_1 \\ x_2 \\ x_3 \\ 1 \end{bmatrix} = \begin{bmatrix} 1 \\ 1 \\ 1 \\ x_4 \end{bmatrix}$$

Using the third diagonal element as pivot gives:

$$\begin{bmatrix} 2 & -1 & -1 & -1 \\ -1 & 2 & 0 & 0 \\ 1 & 0 & 1 & 1 \\ 1 & 0 & 1 & 2 \end{bmatrix} \begin{bmatrix} x_1 \\ x_2 \\ 1 \\ 1 \end{bmatrix} = \begin{bmatrix} 1 \\ 1 \\ x_3 \\ x_4 \end{bmatrix}$$

Using the second diagonal element as pivot gives:

$$\begin{bmatrix} \frac{3}{2} & -\frac{1}{2} & -1 & -1 \\ \frac{1}{2} & \frac{1}{2} & 0 & 0 \\ 1 & 0 & 1 & 1 \\ 1 & 0 & 1 & 2 \end{bmatrix} \begin{bmatrix} x_1 \\ 1 \\ 1 \\ 1 \end{bmatrix} = \begin{bmatrix} 1 \\ x_2 \\ x_3 \\ x_4 \end{bmatrix}$$

Using the first diagonal element as pivot gives the inverse matrix:

$$\mathbf{A}^{-1} = \begin{bmatrix} \frac{2}{3} & \frac{1}{3} & \frac{2}{3} & \frac{2}{3} \\ \frac{1}{3} & \frac{2}{3} & \frac{1}{3} & \frac{1}{3} \\ \frac{2}{3} & \frac{1}{3} & \frac{5}{3} & \frac{5}{3} \\ \frac{2}{3} & \frac{1}{3} & \frac{5}{3} & \frac{8}{3} \end{bmatrix}$$

Hence the solution to the original equation is:

$$\begin{bmatrix} x_1 \\ x_2 \\ x_3 \\ x_4 \end{bmatrix} = \begin{bmatrix} \frac{2}{3} & \frac{1}{3} & \frac{2}{3} & \frac{2}{3} \\ \frac{1}{3} & \frac{2}{3} & \frac{1}{3} & \frac{1}{3} \\ \frac{2}{3} & \frac{1}{3} & \frac{5}{3} & \frac{5}{3} \\ \frac{2}{3} & \frac{1}{3} & \frac{5}{3} & \frac{8}{3} \end{bmatrix} \begin{bmatrix} 1 \\ 1 \\ 1 \\ 1 \end{bmatrix} = \begin{bmatrix} \frac{7}{3} \\ \frac{5}{3} \\ \frac{13}{3} \\ \frac{16}{3} \end{bmatrix}$$

At this point it is worth noting that, although the original coefficient matrix had six zero elements, the inverse matrix is completely full of non-zero elements. This is characteristic of matrix inversion and is why explicit inversion is not a recommended method for solving sparse network equations.

3.2.4 Pivoting

All of the previously described elimination methods fail if the selected pivotal element is zero. For example, the equations

$$a_{12}x_2 + a_{13}x_3 = b_1 \quad (3.15a)$$

$$a_{21}x_1 + a_{22}x_2 + a_{23}x_3 = b_2 \quad (3.15b)$$

$$a_{31}x_1 + a_{32}x_2 + a_{33}x_3 = b_3 \quad (3.15c)$$

cannot be solved using the straightforward sequential process because the first pivot a_{11} is zero. This problem can be solved by interchanging equations 3.15a and 3.15b.

In some special cases, the diagonal elements may have relatively small values which, if used as pivots, could lead to excessive round-off errors. To improve the accuracy of solution in these types of problems, it may be desirable to select suitable pivotal elements. This can be achieved in one of the following ways:

(i) Search the pivotal column in all rows that have not been used as pivotal rows, select the largest element and interchange the row containing this element with the original pivotal row.

(ii) Search the pivotal row including the pivot and all elements to the right of it, select the largest element and interchange the column containing this element with the original pivotal column.

(iii) Search all rows and columns that have not been pivoted, select the largest element and interchange the appropriate columns and appropriate rows to bring this element into the required pivotal position.

(iv) Search all diagonal elements that have not been used as pivots, select the largest element and interchange the appropriate columns and appropriate rows to bring this element into the required pivotal position.

Of these search methods, method (iii) will achieve the greatest accuracy. This method is also the most time-consuming, however, and therefore one of the three compromises may be justified to achieve sufficient accuracy.

Most network problems are characterized by coefficient matrices which contain dominant diagonal elements and are well-conditioned. Therefore, interchanging pivotal elements to reduce round-off errors is not essential in most cases. Many industrial problems having 3000 or more variables, for example, have been solved with adequate accuracy without taking any special measures. However, pivoting can be used to control the degree of sparsity of the coefficient matrix in the elimination process. This can have a significant effect on the number of zeros in the final elimination. Different criteria are required for this purpose, however, and these will be discussed in later chapters.

3.3 Asymmetry and changes in a network

In many design problems it is often necessary to obtain a series of solutions, each subsequent solution differing from the previous one by small changes in the basic data. In power systems operational problems, for example, the network configuration at a given time is fixed except for outage of lines due to overhaul, repair or faults. In order to meet system security requirements, it must be shown that no lines become overloaded when one or more circuits are removed from the basic network. Similarly, it may be necessary to show that a mechanical structure remains stable even when one or more of the supporting elements fails in service. For these and similar analyses, a number of solutions are necessary for small changes in the original network.

This type of analysis can be achieved by continuously changing the coefficient matrix of the basic equations and using one of the standard techniques discussed in the previous sections for the solution of simultaneous equations. However, considerable computation time is spent in re-formulating and re-inverting the whole coefficient matrix for each change in the network.

An alternative method can be used if only one or two changes are made to the network for each solution. This method uses the original inverted coefficient matrix together with a technique for injecting additional nodal quantities which account for the network changes. It saves considerable computation time and involves several distinct concepts. The techniques involved are discussed in detail below.

Consider first the equation

$$\mathbf{AX} = \mathbf{b} \tag{3.16}$$

where $\mathbf{A}$ = the coefficient matrix of the original network

and $\mathbf{b}$ = a column vector containing one element equal to unity and all other elements are equal to zero.

With this definition of **b**, the number of possible vectors representing **b** is equal to the number of equations describing the behaviour of the system. Consider, for example, the following third-order problem:

$$
\begin{aligned}
a_{11}x_1 + a_{12}x_2 + a_{13}x_3 &= 1 \quad &= 0 \quad &= 0\\
a_{21}x_1 + a_{22}x_2 + a_{23}x_3 &= 0 \text{ or } &= 1 \text{ or } &= 0\\
a_{31}x_1 + a_{32}x_2 + a_{33}x_3 &= 0 \quad &= 0 \quad &= 1
\end{aligned}
$$

that is, there are three possible values for **b**. **b** may, therefore, be defined as any one column of a unit matrix having an order equal to the number of equations.

Using the above example and denoting each possible value of **b** by $\mathbf{b}^{(1)}$, $\mathbf{b}^{(2)}$, $\mathbf{b}^{(3)}$ and the corresponding solutions by $\mathbf{X}^{(1)}$, $\mathbf{X}^{(2)}$, $\mathbf{X}^{(3)}$, then:

$$\mathbf{AX}^{(1)} = \mathbf{b}^{(1)},\ \mathbf{AX}^{(2)} = \mathbf{b}^{(2)},\ \mathbf{AX}^{(3)} = \mathbf{b}^{(3)}$$

which can be stated in partitioned matrix form as

$$[\mathbf{A}]\left[\mathbf{X}^{(1)} \mid \mathbf{X}^{(2)} \mid \mathbf{X}^{(3)}\right] = \left[\mathbf{b}^{(1)} \mid \mathbf{b}^{(2)} \mid \mathbf{b}^{(3)}\right]$$

or as:

$$\mathbf{A}\left[\begin{array}{c|c|c} x_1^{(1)} & x_1^{(2)} & x_1^{(3)}\\ x_2^{(1)} & x_2^{(2)} & x_2^{(3)}\\ x_3^{(1)} & x_3^{(2)} & x_3^{(3)} \end{array}\right] = \left[\begin{array}{c|c|c} 1 & 0 & 0\\ 0 & 1 & 0\\ 0 & 0 & 1 \end{array}\right]$$

Now since

$$\mathbf{U} = \mathbf{AA}^{-1}$$

it follows that:

$$\mathbf{A}^{-1} = \left[\begin{array}{c|c|c} x_1^{(1)} & x_1^{(2)} & x_1^{(3)}\\ x_2^{(1)} & x_2^{(2)} & x_2^{(3)}\\ x_3^{(1)} & x_3^{(2)} & x_3^{(3)} \end{array}\right]$$

Therefore the inverse of **A** is given by the matrix formed from the three vectors $\mathbf{X}^{(1)}$, $\mathbf{X}^{(2)}$ and $\mathbf{X}^{(3)}$. Also, it follows that any column of the inverse of **A** can be obtained by solving the network equations with the elements of the right-hand vector **b** set to zero except for the element in the position corresponding to the required column of the inverse matrix. This element must be set to unity.

Although the purpose of the above concept may not be perfectly clear at this stage, the above consideration enables a new solution to be obtained for small changes in the network from a knowledge of the previous solution.

To illustrate the application of this technique, consider the network shown in Fig. 3.1(*a*). The original network can be defined by the equations:

$$\mathbf{AX} = \mathbf{b} \tag{3.17}$$

Consider now the effect of removing the branch between nodes i and j. This gives the network shown in Fig. 3.1(b), which can be defined by:

$$\mathbf{A}^m \mathbf{X}^m = \mathbf{b} \tag{3.18}$$

where $\mathbf{A}^m$ and $\mathbf{X}^m$ are the modified coefficient matrix and the new solution respectively.

The removal of a branch or a change in its value can be represented by the addition of a new branch connected between the same two nodes and having a value (y), either positive or negative, such that the parallel combination of the old and new branches is equivalent to the desired value. In order to represent the removal of a branch, a new branch having a value equal and opposite to the original branch (Y_1) must be connected in parallel as shown in Fig. 3.1(c). This equivalent modified network can still be defined by equation 3.18.

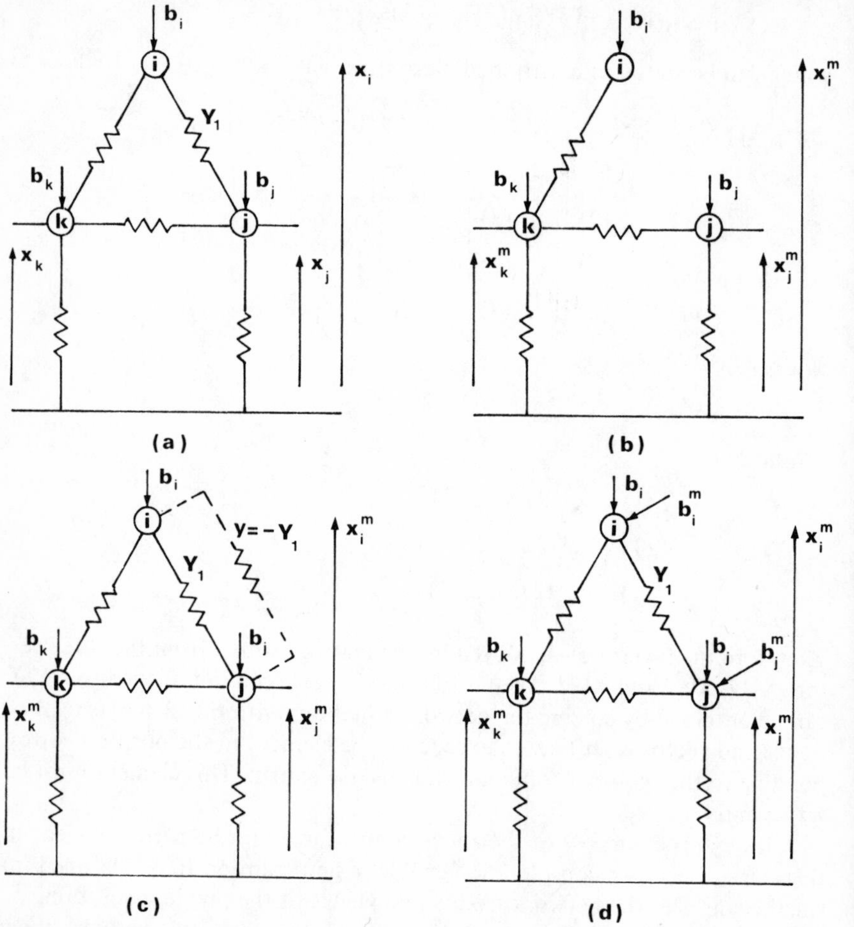

Fig. 3.1 Network modification
(a) original network $\mathbf{AX} = \mathbf{b}$; (b) modified network $\mathbf{A}^m\mathbf{X}^m = \mathbf{b}$; ($c$) equivalent modified network $\mathbf{A}^m\mathbf{X}^m = \mathbf{b}$; ($d$) nodal equivalent modified network $\mathbf{AX}^m = \mathbf{b} + \mathbf{b}^m$

To complete the representation of this network modification, it is now most convenient to replace the additional hypothetical branch by hypothetical injected nodal quantities $\mathbf{b}^m$. This is easily achieved since the 'through' or flow quantity in the new branch is equal to the product of the value of the branch and the potential difference across the branch. The new branch shown in Fig. 3.1(c), for example, can be represented by an injected flow b_i^m at node i and b_j^m (where $b_j^m = -b_i^m$) at node j and where

$$b_i^m = y(x_i^m - x_j^m) \tag{3.19}$$

Therefore, the required modification to the original network as shown in Fig. 3.1(b) can be represented by the nodal equivalent modified network shown in Fig. 3.1(d). This network can be defined by

$$\mathbf{AX}^m = \mathbf{b} + \mathbf{b}^m \tag{3.20}$$

where $\mathbf{A}$ is the coefficient matrix of the original network, since the graph is now unchanged; only the nodal quantities have been altered.

Using the above concepts, it is now possible to derive a simple and computationally efficient method of analysing a network to which several small changes are made.

We will assume that $\mathbf{X}$ has been determined for the original network using the nodal formulation:

$$\mathbf{AX} = \mathbf{b}$$

where $\mathbf{A}$ = original coefficient matrix
and $\mathbf{b}$ = column vector of nodal quantities.

The problem is to find the new value $\mathbf{X}^m$, after one branch in the network has been modified. From equation 3.20,

$$\mathbf{X}^m = \mathbf{A}^{-1}\mathbf{b} + \mathbf{A}^{-1}\mathbf{b}^m \tag{3.21}$$

where $\mathbf{A}^{-1}\mathbf{b}$ = the solutions $\mathbf{X}$ of the original network
and $\mathbf{b}^m$ = a column vector whose elements are all zero except for two.

From equation 3.19, one is equal to $-y(x_i^m - x_j^m)$ and the other is equal to $+y(x_i^m - x_j^m)$, that is,

$$\mathbf{b}^m = -y(x_i^m - x_j^m)\begin{bmatrix} 0 \\ \vdots \\ +1 \\ \vdots \\ -1 \\ \vdots \\ 0 \end{bmatrix} \begin{matrix} \\ \\ i \\ \\ j \\ \\ \\ \end{matrix}$$

Therefore the product $\mathbf{A}^{-1}\mathbf{b}^m$ is also a vector, given by:

$$\mathbf{A}^{-1}\mathbf{b}^{\mathrm{m}} = -y(x_i{}^m - x_j{}^m) \begin{bmatrix} z_{1i} - z_{1j} \\ \vdots \\ z_{ii} - z_{ij} \\ \vdots \\ z_{ji} - z_{jj} \\ \vdots \\ z_{ni} - z_{nj} \end{bmatrix}$$

$$= -y(x_i{}^m - x_j{}^m)\Delta\mathbf{Z} \quad (3.22)$$

where $\Delta\mathbf{Z}$ = the difference between columns i and j of the inverse of the original coefficient matrix.

Substituting equation 3.22 into equation 3.21 gives:

$$\mathbf{X}^{\mathrm{m}} = \mathbf{X} - y(x_i{}^m - x_j{}^m)\Delta\mathbf{Z} \quad (3.23)$$

Writing explicitly the i-th and j-th equations of equation 3.23 gives:

$$x_i{}^m = x_i - y(x_i{}^m - x_j{}^m)(z_{ii} - z_{ij}) \quad (3.24a)$$

$$x_j{}^m = x_j - y(x_i{}^m - x_j{}^m)(z_{ji} - z_{jj}) \quad (3.24b)$$

Subtracting equation 3.24b from equation 3.24a gives:

$$(x_i{}^m - x_j{}^m) = (x_i - x_j) - y(x_i{}^m - x_j{}^m)(z_{ii} + z_{jj} - z_{ij} - z_{ji}) \quad (3.25)$$

Rearranging equation 3.25 gives:

$$(x_i - x_j) = y(x_i{}^m - x_j{}^m)\left(\frac{1}{y} + z_{ii} + z_{jj} - z_{ij} - z_{ji}\right) \quad (3.26)$$

Let $\quad y' = \left(\frac{1}{y} + z_{ii} + z_{jj} - z_{ij} - z_{ji}\right)^{-1}$

then equation 3.26 becomes:

$$y'(x_i - x_j) = y(x_i^m - x_j^m) \quad (3.27)$$

Substituting equation 3.27 into equation 3.23 gives:

$$\mathbf{X}^{\mathrm{m}} = \mathbf{X} - y'(x_i - x_j)\Delta\mathbf{Z} \quad (3.28)$$

Therefore all the new values ($\mathbf{X}^{\mathrm{m}}$) can be found directly from equation 3.28 using the values of $\mathbf{X}$ obtained from the previous solution and the difference ($\Delta\mathbf{Z}$) between columns i and j of the inverse of the original coefficient matrix.

A similar situation exists if two simultaneous modifications are made to the original network. Consider the case when simultaneous modifications are made to the branch between nodes i and j and to the branch between nodes k and l. Equation 3.23 would now be of the form:

$$\mathbf{X}^{\mathrm{m}} = \mathbf{X} - y_{ij}(x_i{}^m - x_j{}^m)\Delta\mathbf{Z}_{\mathrm{ij}} - y_{kl}(x_k{}^m - x_l{}^m)\Delta\mathbf{Z}_{\mathrm{kl}} \quad (3.29)$$

where y_{ij} and y_{kl} are the required modifications to the two branches and $\Delta\mathbf{Z}_{\mathrm{ij}}$ and $\Delta\mathbf{Z}_{\mathrm{kl}}$ are the differences between columns i and j and columns k and l respectively of the original coefficient matrix.

Writing explicitly the i-th, j-th, k-th and l-th equations of equation 3.29 and using the above technique, equation 3.27 becomes:

$$y'_{ij}(x_i - x_j) = y_{ij}(x_i{}^m - x_j{}^m) + y'_{ij}y_{kl}(x_k{}^m - x_l{}^m)\, z'_{kl} \quad (3.30a)$$

$$y'_{kl}(x_k - x_l) = y'_{kl}y_{ij}(x_i{}^m - x_j{}^m)\, z'_{ij} + y_{kl}(x_k{}^m - x_l{}^m) \quad (3.30b)$$

where $$y'_{ij} = \left(\frac{1}{y_{ij}} + z_{ii} + z_{jj} - z_{ij} - z_{ji}\right)^{-1} \text{ and } z'_{kl} = z_{ik} + z_{jl} - z_{il} - z_{jk}$$

$$y'_{kl} = \left(\frac{1}{y_{kl}} + z_{kk} + z_{ll} - z_{kl} - z_{lk}\right)^{-1} \text{ and } z'_{ij} = z_{ki} + z_{lj} - z_{li} - z_{kj}$$

The values $(x_i{}^m - x_j{}^m)$ and $(x_k{}^m - x_l{}^m)$ can be found by solving equations 3.30 simultaneously and substituting these values into equation 3.29. All the new values ($\mathbf{X}^m$) can then be found directly.

It follows from the above examples that, if m simultaneous modifications to the original network are required, then m equations similar to equations 3.27 and 3.30 must be solved. This involves the inversion of an $m \times m$ matrix, which, in general, will be very much smaller than the original $n \times n$ matrix.

In the solution of large sparse equations, most of the computation time is spent in ordering and elimination; the solution for any given right-hand vector can be obtained in a fraction of this time. The above simple procedure avoids repetitive time-consuming elimination when small modifications to the original network are made. This method necessitates only the following simple steps:

(i) the solution of the equation $\mathbf{AX} = \mathbf{b}$ with different $\mathbf{b}$ vectors, these vectors having a unity value in the position which gives the required column of the inverse matrix, all other elements being zero,

(ii) solving a set of simultaneous equations of order m, where m is the number of modifications made at any one time,

(iii) solving a set of equations (e.g. equations 3.23 and 3.28), each involving only one unknown, to find the new potentials at all of the nodes.

This very simple procedure is very efficient computationally when the number of simultaneous modifications are relatively small, and can be repeated for each subsequent modification to the network. For modifications of shunt branches, only the diagonal elements of the coefficient matrix are involved. The process is slightly simplified, therefore, because the potential differences which appear in the above equations are now relative to the zero potential of the reference node. Equation 3.23, for example, becomes:

$$\mathbf{X}^m = \mathbf{X} - yx_i{}^m \,\Delta\mathbf{Z} \qquad (3.31)$$

It is sometimes necessary to analyse networks which have asymmetrical coefficient matrices. These problems can also be solved using matrix inversion techniques or one of the elimination processes. Analysing the network using these methods, however, almost doubles the storage requirements and the computation time. Alternatively, the symmetry of the coefficient matrix $\mathbf{A}$ can be restored by adding terms $y_i x_i{}^m$ to both sides of the equations until the matrix becomes symmetrical. The new equations are therefore equivalent to:

$$\mathbf{AX}^m = \mathbf{b} + \mathbf{b}^m \qquad (3.32)$$

Equation 3.32 can now be solved either by iterative methods, or, if the number of asymmetrical coefficients is relatively small, by the method described above for analysing modified networks. With the latter, step (ii) of the above procedure requires the solution of as many simultaneous equations as there are asymmetrical coefficients.

To illustrate the techniques involved in this method, consider the previous numerical example (page 31) with one element (a_{12}) made asymmetrical. The network equations are therefore:

$$\begin{bmatrix} 3 & -3 & -1 & 0 \\ -1 & 2 & 0 & 0 \\ -1 & 0 & 2 & -1 \\ 0 & 0 & -1 & 1 \end{bmatrix} \begin{bmatrix} x_1{}^m \\ x_2{}^m \\ x_3{}^m \\ x_4{}^m \end{bmatrix} = \begin{bmatrix} 1 \\ 1 \\ 1 \\ 1 \end{bmatrix}$$

Restoring symmetry gives:

$$\begin{bmatrix} 3 & -1 & -1 & 0 \\ -1 & 2 & 0 & 0 \\ -1 & 0 & 2 & -1 \\ 0 & 0 & -1 & 1 \end{bmatrix} \begin{bmatrix} x_1{}^m \\ x_2{}^m \\ x_3{}^m \\ x_4{}^m \end{bmatrix} = \begin{bmatrix} 1 \\ 1 \\ 1 \\ 1 \end{bmatrix} + 2x_2{}^m \begin{bmatrix} 1 \\ 0 \\ 0 \\ 0 \end{bmatrix}$$

which, in general form is:

$$\mathbf{A}\mathbf{X}^{\mathrm{m}} = \mathbf{b} + \mathbf{b}^{\mathrm{m}} \tag{3.33}$$

$$\mathbf{X}^{\mathrm{m}} = \mathbf{A}^{-1}\mathbf{b} + \mathbf{A}^{-1}\mathbf{b}^{\mathrm{m}}$$

where:

$$\mathbf{A}^{-1}\mathbf{b}^{\mathrm{m}} = 2x_2{}^m \; \mathbf{A}^{-1} \begin{bmatrix} 1 \\ 0 \\ 0 \\ 0 \end{bmatrix}$$

From page 39, the first column of A^{-1} gives:

$$\mathbf{A}^{-1}\mathbf{b}^{\mathrm{m}} = 2x_2{}^m \begin{bmatrix} \frac{2}{3} \\ \frac{1}{3} \\ \frac{2}{3} \\ \frac{2}{3} \end{bmatrix}$$

Also, from page 39,

$$A^{-1}b = A^{-1}\begin{bmatrix}1\\1\\1\\1\end{bmatrix} = \begin{bmatrix}\frac{7}{3}\\\frac{5}{3}\\\frac{13}{3}\\\frac{16}{3}\end{bmatrix}$$

Therefore, from equation 3.33,

$$x_1{}^m = \frac{7}{3} + 2x_2{}^m\left(\frac{2}{3}\right)$$

$$x_2{}^m = \frac{5}{3} + 2x_2{}^m\left(\frac{1}{3}\right)$$

$$x_3{}^m = \frac{13}{3} + 2x_2{}^m\left(\frac{2}{3}\right)$$

$$x_4{}^m = \frac{16}{3} + 2x_2{}^m\left(\frac{2}{3}\right)$$

From these equations,

$$x_2{}^m = 5$$

$$x_1{}^m = 9$$

$$x_3{}^m = 11$$

$$x_4{}^m = 12$$

3.4 Iterative methods

3.4.1 Gauss–Seidel iteration

There are many different approximation methods available; these are described in detail in the many books dealing with numerical methods. One of the simplest and most effective methods is the Gauss–Seidel method.

In this method, approximate values are assumed for all but one of the variables, in order to obtain an improved value for the remaining variable. This process is repeated for each variable in turn, with the improved values being used in each subsequent iterative step. Convergence of the iterative process, if obtained, is asymptotic, and therefore the process has to be stopped when the required degree of accuracy has been attained.

This method is best described if the following set of equations are considered:

$$a_{11}x_1 + a_{12}x_2 + a_{13}x_3 = b_1 \quad (3.34a)$$

$$a_{21}x_1 + a_{22}x_2 + a_{23}x_3 = b_2 \quad (3.34b)$$

$$a_{31}x_1 + a_{32}x_2 + a_{33}x_3 = b_3 \quad (3.34c)$$

These equations can be written as:

$$x_1 = \frac{1}{a_{11}} (b_1 - a_{12}x_2 - a_{13}x_3) \quad (3.35a)$$

$$x_2 = \frac{1}{a_{22}} (b_2 - a_{21}x_1 - a_{23}x_3) \quad (3.35b)$$

$$x_3 = \frac{1}{a_{33}} (b_3 - a_{31}x_1 - a_{32}x_2) \quad (3.35c)$$

or, in general terms, as:

$$x_i = \frac{1}{a_{ii}} \left(b_i - \sum_{k \neq i} a_{ik} x_k \right) \quad (3.36)$$

If the n-th approximation of x_k is denoted by $x_k^{(n)}$, then, from equation 3.36, the solution for $x_i^{(n+1)}$ can be obtained from:

$$x_i^{(n+1)} = \frac{1}{a_{ii}} \left(b_i - \sum_{k \neq i} a_{ik} \; x_k^{(n)} \right) \quad (3.37)$$

The original equations 3.34 can be written in matrix form as:

$$\mathbf{AX} = \mathbf{b} \quad (3.38)$$

The coefficient matrix $\mathbf{A}$ can be partitioned into three matrices, such that:

$$\mathbf{A} = \mathbf{L} + \mathbf{D} + \mathbf{H} \quad (3.39)$$

where $\mathbf{L}$ = lower diagonal matrix
$\mathbf{D}$ = diagonal matrix
$\mathbf{H}$ = upper diagonal matrix

For a third order problem as expressed by equations 3.34, these factor matrices are:

$$\mathbf{L} = \begin{bmatrix} \cdot & \cdot & \cdot \\ a_{21} & \cdot & \cdot \\ a_{31} & a_{32} & \cdot \end{bmatrix}, \quad \mathbf{D} = \begin{bmatrix} a_{11} & \cdot & \cdot \\ \cdot & a_{22} & \cdot \\ \cdot & \cdot & a_{33} \end{bmatrix}, \quad \mathbf{H} = \begin{bmatrix} \cdot & a_{12} & a_{13} \\ \cdot & \cdot & a_{23} \\ \cdot & \cdot & \cdot \end{bmatrix}$$

where the dots represent a zero element.

Substituting equation 3.39 into equation 3.38 and rearranging gives:

$$\mathbf{X}^{(n+1)} = \mathbf{D}^{-1} \left[\mathbf{b} - (\mathbf{L} + \mathbf{H})\, \mathbf{X}^{(n)}\right] \tag{3.40}$$

This method is sometimes referred to as the *simultaneous displacement* method because all the new values of $\mathbf{X}[\mathbf{X}^{(n+1)}]$ are evaluated simultaneously from those values $[\mathbf{X}^{(n)}]$ determined during the previous iteration. This method is not used very frequently in practice because, as will be shown in the numerical example at the end of this section (page 50), the number of iterations can become excessive.

The best known and most frequently used alternative method is known as the *successive displacement* method. In this method each new value of x_i is determined sequentially using the most recent evaluation of the other unknown variables x_k. The advantage of this method is that the required number of iterations for a given accuracy can be significantly reduced. For the third-order problem expressed by equation 3.34, this method gives:

$$x_1^{(n+1)} = \frac{1}{a_{11}} (b_1 - a_{12}x_2^{(n)} - a_{13}x_3^{(n)}) \tag{3.41a}$$

$$x_2^{(n+1)} = \frac{1}{a_{22}} (b_2 - a_{21}x_1^{(n+1)} - a_{23}x_3^{(n)}) \tag{3.41b}$$

$$x_3^{(n+1)} = \frac{1}{a_{33}} (b_3 - a_{31}x_1^{(n+1)} - a_{32}x_2^{(n+1)}) \tag{3.41c}$$

Equations 3.41 can be written in matrix form as:

$$\mathbf{X}^{(n+1)} = (\mathbf{D} + \mathbf{L})^{-1} (\mathbf{b} - \mathbf{H}\mathbf{X}^{(n)}) \tag{3.42}$$

To illustrate the techniques of this iterative method consider the previous numerical example shown on page 31. This was

$$\begin{aligned}
3x_1 - x_2 - x_3 \qquad &= 1\\
-x_1 + 2x_2 \qquad &= 1\\
-x_1 \qquad + 2x_3 - x_4 &= 1\\
- x_3 + x_4 &= 1
\end{aligned}$$

From equations 3.35, the iterative equations for the simultaneous displacement method are:

$$x_1^{(n+1)} = \frac{1}{3} (1 + x_2^{(n)} + x_3^{(n)}) \tag{3.43a}$$

$$x_2^{(n+1)} = \frac{1}{2} (1 + x_1^{(n)}) \tag{3.43b}$$

$$x_3^{(n+1)} = \frac{1}{2} (1 + x_1^{(n)} + x_4^{(n)}) \tag{3.43c}$$

$$x_4^{(n+1)} = 1\ (1 + x_3^{(n)}) \tag{3.43d}$$

Initial values must now be estimated for $\mathbf{X}$. Let these be:

$x_1^{(0)} = 2, x_2^{(0)} = 2, x_3^{(0)} = 4, x_4^{(0)} = 5$

The new values $x_i^{(1)}$ can now be calculated using equations 3.43, and the process continued to give the numerical results shown in Table 3.1.

Table 3.1 – Simultaneous displacement method

iteration	x_1	x_2	x_3	x_4
0	2·0000	2·0000	4·0000	5·0000
1	2·3333	1·5000	4·0000	5·0000
2	2·1667	1·1667	4·1667	5·0000
3	2·2778	1·5833	4·0833	5·1667
4	2·2222	1·6389	4·2222	5·0833
5	2·2870	1·6111	4·1528	5·2222
6	2·2546	1·6435	4·2546	5·1528
⋮				
24	2·3294	1·6655	4·3294	5·3243
25	2·3316	1·6647	4·3268	5·3294
26	2·3305	1·6658	4·3305	5·3268
⋮				
40	2·3331	1·6666	4·3331	5·3327
41	2·3332	1·6665	4·3329	5·3331

From Table 3.1, it is seen that 26 iterations were required to achieve a tolerance of 0·005 and 41 iterations to achieve a tolerance of 0·0005.

If the successive displacement method is used and the same initial values for **X** are assumed, then the numerical results shown in Table 3.2 are obtained.

Table 3.2 – Successive displacement method

iteration	x_1	x_2	x_3	x_4
0	2·0000	2·0000	4·0000	5·0000
1	2·3333	1·6667	4·1667	5·1667
2	2·2778	1·6389	4·2222	5·2222
3	2·2870	1·6435	4·2546	5·2546
4	2·2994	1·6497	4·2770	5·2770
5	2·3089	1·6545	4·2930	5·2930
6	2·3158	1·6579	4·3044	5·3044
7	2·3208	1·6604	4·3126	5·3126
8	2·3243	1·6622	4·3184	5·3184
9	2·3269	1·6634	4·3227	5·3227
⋮				
15	2·3325	1·6662	4·3319	5·3319
16	2·3327	1·6664	4·3323	5·3323

From Table 3.2, the advantage of the successive displacement method is clearly evident because, with this method, only 9 iterations are required to achieve a tolerance of 0·005 and 16 iterations for a tolerance of 0·0005.

3.4.2 Relaxation method

The Gauss–Seidel method, described in the previous section, can be regarded as one of the relaxation type processes. In the Gauss–Seidel method each new value of x_i is evaluated explicitly by substituting the previous evaluations of x_k into the appropriate equation for x_i. In the processes normally regarded as relaxation methods, each new value of x_i is estimated using the systematic procedure described below.

If after each iterative step, the most recent values of $x_i^{(n)}$ are substituted into the equations to be solved, a residual $R_i^{(n)}$ will be found which, since the values $x_i^{(n)}$ are approximate, will not be zero. For a third-order problem, these residuals can be defined as:

$$x_1^{(n)} + \frac{a_{12}}{a_{11}} x_2^{(n)} + \frac{a_{13}}{a_{11}} x_3^{(n)} - \frac{b_1}{a_{11}} = R_1^{(n)} \tag{3.44a}$$

$$\frac{a_{21}}{a_{22}} x_1^{(n+1)} + x_2^{(n)} + \frac{a_{23}}{a_{22}} x_3^{(n)} - \frac{b_2}{a_{22}} = R_2^{(n)} \tag{3.44b}$$

$$\frac{a_{31}}{a_{33}} x_1^{(n+1)} + \frac{a_{32}}{a_{33}} x_2^{(n+1)} + x_3^{(n)} - \frac{b_3}{a_{33}} = R_3^{(n)} \tag{3.44c}$$

where $x_i^{(n)}$ is the n-th approximation of the unknown x_i.

The object of the relaxation process is to estimate successive values of $x_i^{(n+1)}$, so that $R_i^{(n)}$ is reduced to zero. This is achieved by selecting the largest residual $R_i^{(n)}$ and estimating $x_i^{(n+1)}$ from:

$$x_i^{(n+1)} = x_i^{(n)} - R_i^{(n)} \tag{3.45}$$

All of the new residuals are then calculated by substituting the value of $x_i^{(n+1)}$, obtained from equation 3.45, into equations 3.44. The process is continued sequentially by choosing the new largest residual.

The main difference between the Gauss–Seidel and the relaxation methods is that with the latter method, the order of the operations is not determined in advance. Instead, the order depends on the values of the current residuals, the largest value normally being chosen for elimination at each stage of the iteration process.

The rate of convergence with this method is good provided that elimination of the largest residual does not affect violently the values of the other residuals. This requirement can sometimes be achieved more easily if a group-relaxation process is used in which several unknowns receive simultaneous increments.

To illustrate the relaxation process, we will again consider the previous numerical example (page 31):

$$\begin{aligned}
3x_1 - x_2 - x_3 \qquad &= 1\\
-x_1 + 2x_2 \qquad\qquad &= 1\\
-x_1 \qquad + 2x_3 - x_4 &= 1\\
- x_3 + x_4 &= 1
\end{aligned}$$

Dividing by the diagonal coefficients and rearranging gives:

$$R_1 = x_1 - \frac{1}{3}x_2 - \frac{1}{3}x_3 - \frac{1}{3} = 0 \tag{3.46a}$$

$$R_2 = -\frac{1}{2}x_1 + x_2 - \frac{1}{2} = 0 \tag{3.46b}$$

$$R_3 = -\frac{1}{2}x_1 + x_3 - \frac{1}{2}x_4 - \frac{1}{2} = 0 \tag{3.46c}$$

$$R_4 = - x_3 + x_4 - 1 = 0 \tag{3.46d}$$

Assuming the initial values of $\mathbf{X}$ to be:

$$x_1^{(0)} = 2, x_2^{(0)} = 2, x_3^{(0)} = 4, x_4^{(0)} = 5$$

gives:

$$R_1^{(0)} = -\ 0{\cdot}3333, R_2^{(0)} = 0{\cdot}5000, R_3^{(0)} = 0{\cdot}0000, R_4^{(0)} = 0{\cdot}0000$$

Choosing the largest residual, $R_2^{(0)}$, the new estimate of x_2 is:

$$\begin{aligned} x_2^{(1)} &= x_2^{(0)} - R_2^{(0)} \\ &= 2{\cdot}0000 - 0{\cdot}5000 \\ &= 1{\cdot}5000 \end{aligned}$$

The other new residuals can now be calculated by substituting $x_2^{(1)}$ into equations 3.46:

$$R_1^{(1)} = -\ 0{\cdot}1667, R_2^{(1)} = 0{\cdot}0000, R_3^{(1)} = 0{\cdot}0000, R_4^{(1)} = 0{\cdot}0000$$

The process can be repeated sequentially until the residuals are less than the required accuracy. For the above example, the numerical results shown in Table 3.3 are obtained.

Table 3.3 – Relaxation method

iteration	x_1	x_2	x_3	x_4	R_1	R_2	R_3	R_4
0	2·0000	2·0000	4·0000	5·0000	−0·3333	0·5000	0·0000	0·0000
1		1·5000			−0·1667	0·0000	0·0000	0·0000
2	2·1667				0·0000	−0·0833	−0·0833	0·0000
3		1·5833			−0·0278	0·0000	−0·0833	0·0000
4			4·0833		−0·0556	0·0000	0·0000	−0·0833
5				5·0833	−0·0556	0·0000	−0·0417	0·0000
6	2·2220				0·0000	−0·0278	−0·0694	0·0000
⋮								
34	2·3233	1·6589	4·3172	5·3110	−0·0021	−0·0027	0·0000	−0·0062
35				5·3172	−0·0021	−0·0027	−0·0031	0·0000
⋮								
58	2·3319	1·6660	4·3313	5·3307	−0·0005	0·0000	0·0000	−0·0006
59				5·3313	−0·0005	0·0000	−0·0003	0·0000

From Tables 3.2 and 3.3, it is seen that the required number of iterations using the relaxation method was much greater than using the successive displacement method to achieve the same degree of accuracy. This difference, however, could be reduced in this problem by using improved relaxation techniques, and in other problems the difference could be reversed.

3.4.3 Acceleration

In many problems, a considerable increase in the rate of convergence can be obtained if an acceleration factor α is used in the iterative process. In practice, the Gauss–Seidel and relaxation methods are rarely used without some form of acceleration. The simplest acceleration comprises a small linear extrapolation of the solution which is applied to every variable at each iterative step. This linear acceleration process takes the form:

$$x_i^{(n+1)} = x_i^{(n)} + \alpha\,(x_i^{(n+1)} - x_i^{(n)}) \tag{3.47}$$

where $x_i^{(n+1)}$ on the right-hand side = latest unaccelerated value of x_i
$x_i^{(n+1)}$ on the left-hand side = accelerated value of x_i

In most problems the value of α is kept constant for all x_i and all iterations. It normally lies in the range:

$$1 < \alpha < 2$$

The optimum value of α depends on the nature of the equations and generally is difficult to determine. It is usual, therefore, to use empirical values. If a correct acceleration factor can be found, the time required for convergence may be reduced by a factor which can vary from two to more than fifty. Acceleration factors which are too large, however, can cause divergence, particularly in the early stages of the iteration, when relatively large differences between successive approximations can exist.

The most practical method of finding the best value of α for a particular set of equations is by trial and error. This can be accomplished by plotting a curve of the number of iterations required for a given degree of accuracy as a function of α. A typical plot is shown in Fig. 3.2, from which the best value of α can be obtained. It is normally found that the value of α is not too critical provided it lies between one and two.

There are other methods of acceleration, such as Aitken's method. Generally these are non-linear and are most effective when monotonic convergence conditions have been reached. The time required to test for these conditions is long, however, and the method is rarely used for the network type of problem.

3.5 Ill-conditioning

If the determinant of the coefficient matrix **A**, det (**A**), is small compared to some of the elements in **A**, then the solution of the equations is very sensitive

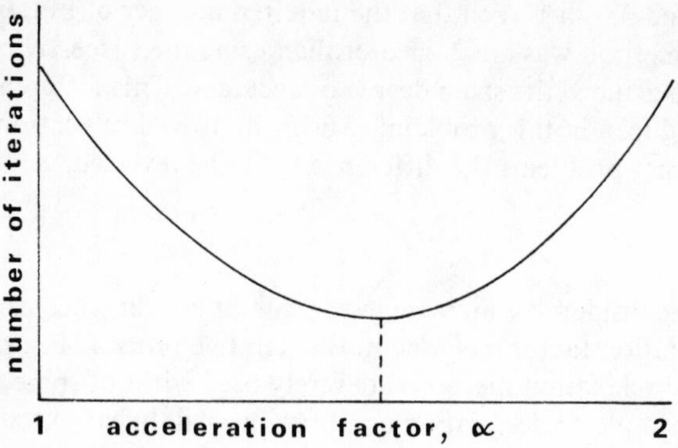

Fig. 3.2 Typical relation between number of iterations and acceleration factor

to round-off errors during computation. The following example represents an extreme case:

$$1000\,x_1 + 2001x_2 = 4003 \tag{3.48a}$$

$$x_1 + \quad 2x_2 = 4 \tag{3.48b}$$

for which $\det(\mathbf{A}) = -1$

The correct solution of equations 3.48 is:

$x_1 = -2$

$x_2 = \ \ 3$

If the coefficient of x_2 in equation 3.48a is changed by only −0·1 per cent to 1999, which could represent a rounding-off error, the solution of equations 3.48 becomes:

$x_1 = 10$

$x_2 = -3$

and if the same coefficient is changed by +0·1 per cent to 2003, the solution becomes:

$x_1 = 2$

$x_2 = 1$

A comparison of these solutions shows that only a very small change in one of the coefficients can produce a very significant error in the computed solution.

The set of equations in which the problem occurs are said to be *ill-conditioned.* The problem can be illustrated for a second-order problem by comparing the variation of x_2 with x_1 as shown in Fig. 3.3.

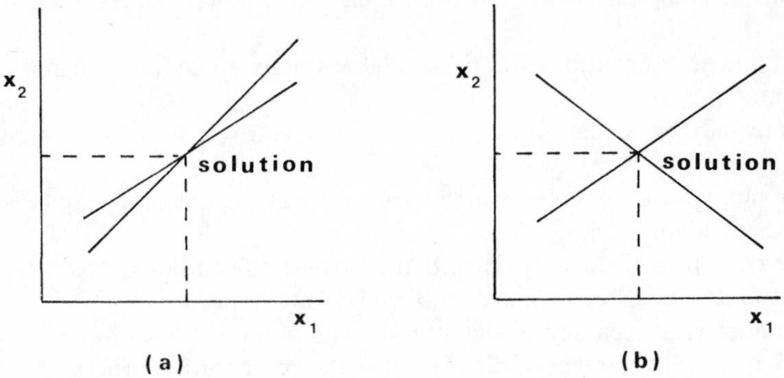

Fig. 3.3 Graphical representation of ill- and well-conditioned equations. (*a*) ill-conditioned; (*b*) well-conditioned

Each straight line in Fig. 3.3 represents one of the equations. In the ill-conditioned problem shown in Fig. 3.3(*a*), the angle between the two lines is small and therefore the point of intersection, which represents the solution, is difficult to calculate accurately. This problem does not occur in the well-conditioned problem shown in Fig. 3.3(*b*).

To determine numerically a measure of ill-conditioning is a long process and it is often simpler in practice to recognize the effect by certain symptoms of the problem. If some of the diagonal elements are small compared with some off-diagonal terms, solution of the equations may be difficult. This problem of ill-conditioning is more apparent with iterative techniques than with direct methods.

Ill-conditioned equations can be created for one of two reasons. Firstly, the physical nature of the system may be such that it is itself an ill-conditioned problem. In this case the equations describing its behaviour will be ill-conditioned irrespective of the method chosen to model the system. There is little that can be done except to ensure that the largest elements are used as pivots, that the calculations are carried out with sufficient digits to minimize rounding-off errors, and that the solution is carefully checked for accuracy. Secondly, the system itself may be well-conditioned but the modelling method may create a set of ill-conditioned equations. In this case it is usually possible to re-formulate the equations so as to improve the conditioning.

3.6 Choice of method

3.6.1 Factors affecting choice

Standard programs for the solution of a set of simultaneous linear equations exist with every organized digital computing service. A number of factors must be considered before choosing which method to use in order to achieve a reasonable computational efficiency. Some of the most important factors are:

(i) the number of equations and the available storage space in the digital computer

(ii) the form of the equations, for example, symmetrical, diagonal, number of zero elements

(iii) the number of unknowns which need to be solved; this may vary between one and all of them

(iv) the condition of the equations, that is well- or ill-conditioned

(v) the number of times a similar problem is likely to occur

(vi) the numerical accuracy (which can depend on the accuracy of the coefficients), the number of digits used in the computation process, the required accuracy and the sensitivity of the problem to errors.

3.6.2 Comparison of methods

The iterative methods are useful for solving a large number of equations which contain a high proportion of zero elements. They are very economical in computer storage since they require only the space of the original coefficients. They are also relatively easy to program. Until fairly recently, they were virtually the only practical method of solving very large problems. They are not very suitable for ill-conditioned equations, however, and can evaluate the solution only to within a certain degree of accuracy. Also, their rate of convergence is dependent on choosing a suitable acceleration factor and they may, in very unfortunate cases, actually diverge from the solution.

The direct methods have complete general applicability. They can determine the inverse matrix in explicit or factored form in a finite and predeterminable number of arithmetic operations. Advantage can be taken of sparsity to reduce the amount of computer storage space required and the length of the computational time. To enable the efficient use of sparsity in order to minimize the number of non-zero elements during the elimination process, however, it is necessary to employ special techniques and very skilful programming. It is only fairly recently that these techniques and programs have been sufficiently developed to make direct methods generally preferable to the iterative methods. Commercial programs are now available, based on direct methods employing advanced sparsity techniques, which can solve sparse problems of 1000 variables in about 1s on a CDC 7600 and which have also solved problems of 5000 variables without any difficulties. The concepts of these techniques are explained in the following chapters.

Bibliography

Berezin, O. M. & Zhidkov, N. P.: *Computing Methods*, vol. II, chap. 6. Pergamon Press, 1965.

Bertelé, V. & Brioschi, F.: On the theory of the elimination process. *J. math. Anal. Applic.* **35**, 48–57, 1971.
Brameller, A.: Efficient multiple solutions for changes in a network using sparsity techniques. *Proc. Instn elect. Engrs* **120**, 607–8, 1973.
Brameller, A. & Allan, R. N.: The role of sparsity in the analysis of large systems. *Comput. Aided Des.* **6**, 159–68, 1974.
Chen, Y. T. & Tewarson, R. P.: On the optimal choice of pivots for the Gaussian elimination. *Computing* **9**, 245–50, 1972.
Crandal, S. H.: *Engineering Analysis.* McGraw-Hill, 1956.
Douglas, A.: Examples concerning efficient strategies for Gaussian elimination. *Computing* **8**, 382–94, 1971.
Evans, D. J.: New iterative procedures for the solution of sparse systems of linear difference equations. In *Sparse Matrices and their Applications,* pp. 89–100. Plenum Press, 1972.
Forsythe, G. E. & Moler, C. B.: *Computer Solution of Linear Algebraic Systems.* Prentice-Hall, 1967.
Fox, L.: *Introduction to Numerical Linear Algebra.* Clarendon Press, 1965.
Gerald, C. F.: *Applied Numerical Analysis.* Addison-Wesley, 1970.
Hildebrand, F. B.: *Introduction to Numerical Analysis.* McGraw-Hill, 1956.
McCracken, D. D. & Dorn, W. S.: *Numerical Methods and Fortran Programming.* John Wiley, 1968.
Nielson, K. L.: *Methods in Numerical Analysis.* Macmillan, 1968.
Ralston, A.: *A First Course in Numerical Analysis.* McGraw-Hill, 1965.
Stagg, G. W. & El-Abiad, A. H.: *Computer Methods in Power System Analysis.* McGraw-Hill, 1968.
Tewarson, R. P.: The Crout reduction for sparse matrices. *Comput. J.* **12**, 158–9, 1969.
Tewarson, R. P.: The Gaussian elimination and sparse systems. In *Sparse Matrix Proceedings,* pp. 35–42. IBM, 1969.
Varga, R. A.: *Matrix Iterative Analysis.* Prentice-Hall, 1962.
Walsh, J.: Direct and indirect methods. In *Large Sparse Sets of Linear Equations,* pp. 41–56. Academic Press, 1971.
Westlake, J. R.: *A Handbook of Numerical Matrix Inversion and Solution of Linear Equations.* John Wiley, 1968.
Wilkinson, J. H.: *The Algebraic Eigenvalue Problem.* Oxford University Press, 1965.

4
Matrix Factorization

4.1 Introduction

The analysis of a large system using the network approach frequently necessitates the solution of hundreds and maybe thousands of simultaneous equations having the form **AX** = **b**. Furthermore, several solutions are often required with the same coefficient matrix **A** but with a series of different **b** vectors. Such equations can be solved using any of the conventional and elementary methods described in chapter 3. One possibility is direct matrix inversion, but, as discussed in chapter 3, this requires n^2 storage locations for the coefficients and about n^3 arithmetic operations for the solution of n simultaneous linear equations. Even if the matrix **A** is very sparse, its inverse is completely full and therefore this method is normally a very inefficient technique for solving a large number of equations.

An alternative method is Gauss elimination. As discussed in chapter 3, this reduces the number of arithmetic operations to about $\frac{n^3}{3}$, but requires an indeterminate number of storage locations, which will be less than or equal to that required by direct inversion methods. It is therefore significantly better than direct inversion, but requires a systematic form of logic to achieve an efficient computer program. This is achieved using one of the various modifications of the basic Gauss elimination technique, which are generally known as matrix factorization methods. These methods use Gauss elimination to obtain the inverse of the coefficient matrix implicitly as the product of several factor matrices. They do not in themselves improve on the storage requirements or the number of arithmetic operations needed using Gauss elimination. However, because of their systematized logic, they lend themselves to numerical techniques and computer programming, which, when sparsity techniques are included, can drastically reduce both the number of operations and storage requirements.

There are many possible factorization methods and adaptations. In this chapter some of the most important ones which have been developed over the past years and which have been found to be most useful are described. It should be noted, however, that, although the form and manipulation may seem to be complex, these methods are basically very simple and are only systematic extensions of the basic Gauss elimination technique.

4.2 Product form of the inverse

The product form of the inverse is one of the most straightforward of all factorization methods. It is not frequently used for analysing sparse networks as there are more important and efficient methods available. It has, however, an important application in the solution of linear programming problems. It is also a formalized method of performing Gauss elimination and illustrates clearly how a series of factor matrices may be used to solve a set of simultaneous equations and how the inverse of the coefficient matrix may be obtained implicitly.

Consider the equation

$$\mathbf{AX} = \mathbf{b} \tag{4.1}$$

the solution of which is:

$$\mathbf{X} = \mathbf{A}^{-1}\mathbf{b} \tag{4.2}$$

In the product form of the inverse, the matrix $\mathbf{A}^{-1}$ is not calculated explicitly but is obtained by multiplying n factor matrices, i.e.

$$\mathbf{A}^{-1} = \mathbf{T}_n \ldots \mathbf{T}_3 \mathbf{T}_2 \mathbf{T}_1 \tag{4.3}$$

The steps required to achieve this result (equation 4.3) are best illustrated by considering the following third-order problem:

$$\begin{bmatrix} a_{11} & a_{12} & a_{13} \\ a_{21} & a_{22} & a_{23} \\ a_{31} & a_{32} & a_{33} \end{bmatrix} \begin{bmatrix} x_1 \\ x_2 \\ x_3 \end{bmatrix} = \begin{bmatrix} b_1 \\ b_2 \\ b_3 \end{bmatrix}$$

In the Gauss elimination technique, the elements below the diagonal element of the first column are eliminated by using the first diagonal coefficient as a pivot. This can be achieved by dividing the first row by a_{11}, multiplying the new row by a_{21} and a_{31} respectively and subtracting these results from the second and third row respectively. Instead of using this technique, the same series of operations may be achieved by pre-multiplying the coefficient matrix $\mathbf{A}$ by a transformation matrix $\mathbf{T}_1$, where:

$$\mathbf{T}_1 = \begin{bmatrix} \dfrac{1}{a_{11}} & \cdot & \cdot \\ -\dfrac{a_{21}}{a_{11}} & 1 & \cdot \\ -\dfrac{a_{31}}{a_{11}} & \cdot & 1 \end{bmatrix}$$

This operation gives a new matrix $\mathbf{A}^{(1)} = \mathbf{T}_1 \mathbf{A}$, where:

$$\mathbf{A}^{(1)} = \begin{bmatrix} 1 & a_{12}^{(1)} & a_{13}^{(1)} \\ \cdot & a_{22}^{(1)} & a_{23}^{(1)} \\ \cdot & a_{32}^{(1)} & a_{33}^{(1)} \end{bmatrix}$$

The elements $a_{ij}^{(1)}$ of $\mathbf{A}^{(1)}$ are obtained by the method used in the Gauss elimination method, discussed in chapter 3, i.e.

$$a_{ij}^{(1)} = a_{ij} - \frac{a_{i1} a_{1j}}{a_{11}}$$

where $i = 1, 2 \ldots n$
$j = 2, \ldots n$

Therefore, equation 4.1 is transformed into a related set of equations which can be expressed as:

$$\mathbf{A}^{(1)}\mathbf{X} = \mathbf{T}_1 \mathbf{A}\mathbf{X} = \mathbf{T}_1 \mathbf{b} \tag{4.4}$$

This process can be continued using the second diagonal element of the new coefficient matrix $\mathbf{A}^{(1)}$ as a pivot. Using the same technique, the off-diagonal elements of the second column of $\mathbf{A}^{(1)}$ can be reduced to zero and the diagonal element made unity by pre-multiplying the matrix $\mathbf{A}^{(1)}$ by a transformation matrix $\mathbf{T}_2$, where:

$$\mathbf{T}_2 = \begin{bmatrix} 1 & -\dfrac{a_{12}^{(1)}}{a_{22}^{(1)}} & \cdot \\ \cdot & \dfrac{1}{a_{22}^{(1)}} & \cdot \\ \cdot & -\dfrac{a_{32}^{(1)}}{a_{22}^{(1)}} & 1 \end{bmatrix}$$

This operation gives a new matrix $\mathbf{A}^{(2)} = \mathbf{T}_2 \mathbf{A}^{(1)} = \mathbf{T}_2 \mathbf{T}_1 \mathbf{A}$, where

$$\mathbf{A}^{(2)} = \begin{bmatrix} 1 & \cdot & a_{13}^{(2)} \\ \cdot & 1 & a_{23}^{(2)} \\ \cdot & \cdot & a_{33}^{(2)} \end{bmatrix}$$

and equation 4.1 is now transformed to:

$$\mathbf{A}^{(2)}\mathbf{X} = \mathbf{T}_2 \mathbf{A}^{(1)}\mathbf{X} = \mathbf{T}_2 \mathbf{T}_1 \mathbf{A}\mathbf{X} = \mathbf{T}_2 \mathbf{T}_1 \mathbf{b} \tag{4.5}$$

If this transformation process is continued, then for a n-th order problem, equation 4.1 becomes:

$$\mathbf{A}^{(n)}\mathbf{X} = \mathbf{T}_n \ldots \mathbf{T}_2\mathbf{T}_1\mathbf{A}\mathbf{X} = \mathbf{T}_n \ldots \mathbf{T}_2\mathbf{T}_1\mathbf{b} \tag{4.6}$$

But $\mathbf{A}^{(n)}$ has been reduced sequentially to a unit matrix.
Therefore, equation 4.6 is:

$$\mathbf{X} = \mathbf{T}_n \ldots \mathbf{T}_2\mathbf{T}_1\mathbf{b} \tag{4.7}$$

and by comparing equations 4.2 and 4.7

$$\mathbf{A}^{-1} = \mathbf{T}_n \ldots \mathbf{T}_2\mathbf{T}_1 \tag{4.8}$$

Therefore, from equation 4.8, it is seen that this transformation process enables the inverse of the original matrix $\mathbf{A}$ to be obtained implicitly as the product of n factor or transformation matrices.

Each transformation matrix $\mathbf{T}_i (i = 1, \ldots n)$ is a unit matrix except for its i-th column. Therefore, in digital computer solutions, only this i-th column need be stored; all other elements of the matrix are known implicitly. In general sparse network problems, the i-th column of $\mathbf{T}_i$ will also contain a large proportion of zero elements.

As an example of this technique, we will again consider the same numerical example used in chapter 3. The matrix form of these equations were:

$$\begin{bmatrix} 3 & -1 & -1 & \cdot \\ -1 & 2 & \cdot & \cdot \\ -1 & \cdot & 2 & -1 \\ \cdot & \cdot & -1 & 1 \end{bmatrix} \begin{bmatrix} x_1 \\ x_2 \\ x_3 \\ x_4 \end{bmatrix} = \begin{bmatrix} 1 \\ 1 \\ 1 \\ 1 \end{bmatrix}$$

The first transformation matrix $\mathbf{T}_1$ is obtained from the first column of the coefficient matrix, and is:

$$\mathbf{T}_1 = \begin{bmatrix} \frac{1}{3} & \cdot & \cdot & \cdot \\ \frac{1}{3} & 1 & \cdot & \cdot \\ \frac{1}{3} & \cdot & 1 & \cdot \\ \cdot & \cdot & \cdot & 1 \end{bmatrix}$$

The reduced coefficient matrix $\mathbf{A}^{(1)}$ is obtained by using Gauss elimination and is:

$$\mathbf{A}^{(1)} = \mathbf{T}_1\,\mathbf{A} = \begin{bmatrix} 1 & -\frac{1}{3} & -\frac{1}{3} & \cdot \\ \cdot & \frac{5}{3} & \frac{1}{3} & \cdot \\ \cdot & -\frac{1}{3} & \frac{5}{3} & -1 \\ \cdot & \cdot & -1 & 1 \end{bmatrix}$$

Similarly, the second transformation matrix and the corresponding reduced coefficient matrix are:

$$\mathbf{T}_2 = \begin{bmatrix} 1 & \frac{1}{5} & \cdot & \cdot \\ \cdot & \frac{3}{5} & \cdot & \cdot \\ \cdot & \frac{1}{5} & 1 & \cdot \\ \cdot & \cdot & \cdot & 1 \end{bmatrix} \qquad \mathbf{A}^{(2)} = \mathbf{T}_2\,\mathbf{A}^{(1)} = \begin{bmatrix} 1 & \cdot & -\frac{2}{5} & \cdot \\ \cdot & 1 & -\frac{1}{5} & \cdot \\ \cdot & \cdot & \frac{8}{5} & -1 \\ \cdot & \cdot & -1 & 1 \end{bmatrix}$$

The third transformation matrix and the corresponding reduced coefficient matrix are:

$$\mathbf{T}_3 = \begin{bmatrix} 1 & \cdot & \frac{1}{4} & \cdot \\ \cdot & 1 & \frac{1}{8} & \cdot \\ \cdot & \cdot & \frac{5}{8} & \cdot \\ \cdot & \cdot & \frac{5}{8} & \cdot \end{bmatrix} \qquad \mathbf{A}^{(3)} = \mathbf{T}_3\,\mathbf{A}^{(2)} = \begin{bmatrix} 1 & \cdot & \cdot & -\frac{1}{4} \\ \cdot & 1 & \cdot & -\frac{1}{8} \\ \cdot & \cdot & 1 & -\frac{5}{8} \\ \cdot & \cdot & \cdot & \frac{3}{8} \end{bmatrix}$$

The final transformation matrix and the corresponding reduced coefficient matrix are:

$$\mathbf{T}_4 = \begin{bmatrix} 1 & \cdot & \cdot & \frac{2}{3} \\ \cdot & 1 & \cdot & \frac{1}{3} \\ \cdot & \cdot & 1 & \frac{5}{3} \\ \cdot & \cdot & \cdot & \frac{8}{3} \end{bmatrix} \qquad \mathbf{A}^{(4)} = \mathbf{T}_4\,\mathbf{A}^{(3)} = \begin{bmatrix} 1 & \cdot & \cdot & \cdot \\ \cdot & 1 & \cdot & \cdot \\ \cdot & \cdot & 1 & \cdot \\ \cdot & \cdot & \cdot & 1 \end{bmatrix}$$

Using equation 4.7, the solution to the original equations can now be calculated from:

$$\mathbf{X} = \mathbf{T}_4\,\mathbf{T}_3\,\mathbf{T}_2\,\mathbf{T}_1\,\mathbf{b}$$

which in matrix form is:

$$\begin{bmatrix} x_1 \\ x_2 \\ x_3 \\ x_4 \end{bmatrix} = \begin{bmatrix} 1 & \cdot & \cdot & \frac{2}{3} \\ \cdot & 1 & \cdot & \frac{1}{3} \\ \cdot & \cdot & 1 & \frac{5}{3} \\ \cdot & \cdot & \cdot & \frac{8}{3} \end{bmatrix} \begin{bmatrix} 1 & \cdot & \frac{1}{4} & \cdot \\ \cdot & 1 & \frac{1}{8} & \cdot \\ \cdot & \cdot & \frac{5}{8} & \cdot \\ \cdot & \cdot & \frac{5}{8} & 1 \end{bmatrix} \begin{bmatrix} 1 & \frac{1}{5} & \cdot & \cdot \\ \cdot & \frac{3}{5} & \cdot & \cdot \\ \cdot & \frac{1}{5} & 1 & \cdot \\ \cdot & \cdot & \cdot & 1 \end{bmatrix} \begin{bmatrix} \frac{1}{3} & \cdot & \cdot & \cdot \\ \frac{1}{3} & 1 & \cdot & \cdot \\ \frac{1}{3} & \cdot & 1 & \cdot \\ \cdot & \cdot & \cdot & 1 \end{bmatrix} \begin{bmatrix} 1 \\ 1 \\ 1 \\ 1 \end{bmatrix} = \begin{bmatrix} \frac{1}{3} \\ \frac{5}{3} \\ \frac{13}{3} \\ \frac{16}{3} \end{bmatrix}$$

It is clearly evident from this numerical example that only the *i*-th column of the *i*-th transformation matrix need be stored and that each new reduced coefficient matrix can be overwritten on the preceding one.

4.3 Triangulation of matrices

Another effective and most widely used method of manipulating coefficient matrices to solve simultaneous linear equations is that associated with *triangulation* of matrices or *triangular decomposition.* These methods factorize the coefficient matrix into their triangular form, on which several important and efficient modern techniques are based. The two methods which are discussed in this book are generally known as the LH (or sometimes LU) and LDH (or LDU) methods.

4.3.1 LH factorization

The LH method of factorization consists of expressing the coefficient matrix **A** as the product of two factor matrices, such that:

$$\mathbf{A} = \mathbf{LH} \tag{4.9}$$

where **L** = a lower triangular matrix
H = a higher (or upper) triangular matrix which has unity elements on its diagonal.

The ability to factorize in this way is a fundamental property of any square matrix; for a third-order problem these factor matrices are:

$$\begin{bmatrix} a_{11} & a_{12} & a_{13} \\ a_{21} & a_{22} & a_{23} \\ a_{31} & a_{32} & a_{33} \end{bmatrix} = \begin{bmatrix} l_{11} & \cdot & \cdot \\ l_{21} & l_{22} & \cdot \\ l_{31} & l_{32} & l_{33} \end{bmatrix} \begin{bmatrix} 1 & h_{12} & h_{13} \\ \cdot & 1 & h_{23} \\ \cdot & \cdot & 1 \end{bmatrix}$$

Multiplying **L** and **H** and equating the elements of this matrix with the corresponding elements of **A** gives:

$$\begin{array}{lll} a_{11} = l_{11} & a_{12} = l_{11}h_{12} & a_{13} = l_{11}h_{13} \\ a_{21} = l_{21} & a_{22} = l_{21}h_{12} + l_{22} & a_{23} = l_{21}h_{13} + l_{22}h_{23} \\ a_{31} = l_{31} & a_{32} = l_{31}h_{12} + l_{32} & a_{33} = l_{31}h_{13} + l_{32}h_{23} + l_{33} \end{array}$$

and rearranging gives

$$\begin{array}{lll} l_{11} = a_{11} & h_{12} = \dfrac{a_{12}}{l_{11}} & h_{13} = \dfrac{a_{13}}{l_{11}} \\ l_{21} = a_{21} & l_{22} = a_{22} - l_{21}h_{12} & h_{23} = \dfrac{1}{l_{22}}(a_{23} - l_{21}h_{13}) \\ l_{31} = a_{31} & l_{32} = a_{32} - l_{31}h_{12} & l_{33} = a_{33} - l_{31}h_{13} - l_{32}h_{23} \end{array}$$

From these equations it is evident that there are unique values of l and h which give the original elements of **A**.

If the set of simultaneous linear equations to be solved are written in matrix form as:

$$\mathbf{AX} = \mathbf{b} \tag{4.10}$$

then substituting equation 4.9 into equation 4.10 gives:

$$\mathbf{LHX} = \mathbf{b} \tag{4.11}$$

Letting $\mathbf{HX} = \mathbf{Y}$ (4.12)

then from equations 4.11 and 4.12,

$$\mathbf{LY} = \mathbf{b} \tag{4.13}$$

Writing explicitly equations 4.12 and 4.13 for a third-order problem gives:

$$\begin{aligned} x_1 + h_{12}x_2 + h_{13}x_3 &= y_1 & l_{11}y_1 &= b_1 \\ x_2 + h_{23}x_3 &= y_2 & l_{21}y_1 + l_{22}y_2 &= b_2 \\ x_3 &= y_3 & l_{31}y_1 + l_{32}y_2 + l_{33}y_3 &= b_3 \end{aligned}$$

Since **L** is a lower triangular matrix, **Y** can be found from **L** and **b** by forward substitution, and since **H** is an upper triangular matrix, the unknown vector **X** can be found from **H** and **Y** by backward substitution. For a third-order problem, this technique gives the values of **Y** and **X** as:

$$\begin{aligned} y_1 &= \frac{b_1}{l_{11}} & x_3 &= y_3 \\ y_2 &= \frac{1}{l_{22}}(b_2 - l_{21}y_1) & x_2 &= y_2 - h_{23}x_3 \\ y_3 &= \frac{1}{l_{33}}(b_3 - l_{31}y_1 - l_{32}y_2) & x_1 &= y_1 - h_{12}x_2 - h_{13}x_3 \end{aligned}$$

The problem of this technique is therefore to determine **L** and **H**; this can readily be achieved using a systematic modification of the basic Gauss elimination process.

Consider the third-order coefficient matrix:

$$\mathbf{A} = \begin{bmatrix} a_{11} & a_{12} & a_{13} \\ a_{21} & a_{22} & a_{23} \\ a_{31} & a_{32} & a_{33} \end{bmatrix}$$

In the Gauss elimination process, the elements below the diagonal element of the first column are eliminated by using the first diagonal element as a pivot. This can be achieved by dividing the first row by a_{11}, multiplying the new row by a_{21} and a_{31} respectively and subtracting these results from the second and third row respectively. Instead of using this technique, the same series of operations may be achieved by pre-multiplying the coefficient matrix **A** by a transformation matrix $\mathbf{L}_1$, where

$$\mathbf{L}_1 = \begin{bmatrix} \frac{1}{a_{11}} & \cdot & \cdot \\ -\frac{a_{21}}{a_{11}} & 1 & \cdot \\ -\frac{a_{31}}{a_{11}} & \cdot & 1 \end{bmatrix}$$

(L_1 is, incidentally, identical to the transformation matrix T_1 discussed on page 59).

This operation gives a new coefficient matrix $\mathbf{A}^{(1)}$, where:

$$\mathbf{A}^{(1)} = \mathbf{L}_1 \mathbf{A}$$

$$\text{or} \quad \mathbf{L}_1^{-1} \mathbf{A}^{(1)} = \mathbf{A} \tag{4.14}$$

where:

$$\mathbf{L}_1^{-1} = \begin{bmatrix} a_{11} & \cdot & \cdot \\ a_{21} & 1 & \cdot \\ a_{31} & \cdot & 1 \end{bmatrix} \quad \text{and } \mathbf{A}^{(1)} = \begin{bmatrix} 1 & a_{12}^{(1)} & a_{13}^{(1)} \\ \cdot & a_{22}^{(1)} & a_{23}^{(1)} \\ \cdot & a_{32}^{(1)} & a_{33}^{(1)} \end{bmatrix}$$

The elements $a_{ij}^{(1)}$ of $\mathbf{A}^{(1)}$ are obtained by the method used in the Gauss elimination method, discussed in chapter 3, i.e.

$$a_{ij}^{(1)} = a_{ij} - \frac{a_{i1}a_{1j}}{a_{11}}$$

$$\text{where } i = 1, 2, \ldots n$$
$$j = \quad 2, \ldots n$$

This process can be continued using the second diagonal element of the new coefficient matrix $\mathbf{A}^{(1)}$ as a pivot. Using the same technique, the elements below the second diagonal of $\mathbf{A}^{(1)}$ can be eliminated by pre-multiplying the matrix $\mathbf{A}^{(1)}$ by a lower triangular transformation matrix $\mathbf{L}_2$, where:

$$\mathbf{L}_2 = \begin{bmatrix} 1 & \cdot & \cdot \\ \cdot & \dfrac{1}{a_{22}^{(1)}} & \cdot \\ \cdot & -\dfrac{a_{32}^{(1)}}{a_{22}^{(1)}} & \cdot \end{bmatrix}$$

This operation gives a new matrix $\mathbf{A}^{(2)}$, where:

$$\mathbf{A}^{(2)} = \mathbf{L}_2 \mathbf{A}^{(1)}$$

$$\text{or} \quad \mathbf{A}^{(1)} = \mathbf{L}_2^{-1} \mathbf{A}^{(2)} \tag{4.15}$$

From equations 4.14 and 4.15,

$$\mathbf{A} = \mathbf{L}_1^{-1} \mathbf{L}_2^{-1} \mathbf{A}^{(2)}$$

Continuing this process for each diagonal element in turn gives, for an n-th order problem:

$$\mathbf{A} = \mathbf{L}_1^{-1} \mathbf{L}_2^{-1} \ldots \mathbf{L}_n^{-1} \mathbf{A}^{(n)} \tag{4.16}$$

where $\mathbf{A}^{(n)}$ will be the upper triangular matrix $\mathbf{H}$, which, for a third-order problem, is:

$$\mathbf{A}^{(n)} = \mathbf{H} = \begin{bmatrix} 1 & a_{12}^{(1)} & a_{13}^{(2)} \\ \cdot & 1 & a_{23}^{(2)} \\ \cdot & \cdot & 1 \end{bmatrix}$$

Also, since the factors $\mathbf{L}_1 \ldots \mathbf{L}_n$ are lower triangular matrices, the product $\mathbf{L}$ of these individual matrices will be a lower triangular matrix, where:

$$\mathbf{L} = \mathbf{L}_1^{-1}\mathbf{L}_2^{-1} \ldots \mathbf{L}_n^{-1}$$

Therefore, for the third-order problem:

$$\mathbf{L} = \begin{bmatrix} a_{11} & \cdot & \cdot \\ a_{21} & 1 & \cdot \\ a_{31} & \cdot & 1 \end{bmatrix} \begin{bmatrix} 1 & \cdot & \cdot \\ \cdot & a_{22}^{(1)} & \cdot \\ \cdot & a_{32}^{(1)} & \cdot \end{bmatrix} \begin{bmatrix} 1 & \cdot & \cdot \\ \cdot & 1 & \cdot \\ \cdot & \cdot & a_{33}^{(2)} \end{bmatrix} = \begin{bmatrix} a_{11} & \cdot & \cdot \\ a_{21} & a_{22}^{(1)} & \cdot \\ a_{31} & a_{32}^{(1)} & a_{33}^{(2)} \end{bmatrix}$$

Since the diagonal elements of $\mathbf{H}$ are always unity, they do not need to be stored explicitly when using a digital computer. The two triangular matrices, therefore, can be combined for storage purposes, the diagonal elements of the combination being those of $\mathbf{L}$.

As an example of this technique consider the previous numerical example, which was:

$$\begin{bmatrix} 3 & -1 & -1 & \cdot \\ -1 & 2 & \cdot & \cdot \\ -1 & \cdot & 2 & -1 \\ \cdot & \cdot & -1 & 1 \end{bmatrix} \begin{bmatrix} x_1 \\ x_2 \\ x_3 \\ x_4 \end{bmatrix} = \begin{bmatrix} 1 \\ 1 \\ 1 \\ 1 \end{bmatrix}$$

Using the Gauss elimination process and overwriting the elements of $\mathbf{A}$ with the elements of the new reduced coefficient matrix and the elements of $\mathbf{L}$ and $\mathbf{H}$ at each reduction step gives:

the first reduction step as:

$$\left[\begin{array}{c|ccc} 3 & -\frac{1}{3} & -\frac{1}{3} & \cdot \\ \hline -1 & \frac{5}{3} & -\frac{1}{3} & \cdot \\ -1 & -\frac{1}{3} & \frac{5}{3} & -1 \\ \cdot & \cdot & -1 & 1 \end{array}\right]$$

the second reduction step as:

$$\begin{bmatrix} 3 & -\frac{1}{3} & -\frac{1}{3} & \cdot \\ -1 & \frac{5}{3} & -\frac{1}{5} & \cdot \\ -1 & -\frac{1}{3} & \frac{8}{5} & -1 \\ \cdot & \cdot & -1 & 1 \end{bmatrix}$$

the third reduction step as:

$$\begin{bmatrix} 3 & -\frac{1}{3} & -\frac{1}{3} & \cdot \\ -1 & \frac{5}{3} & -\frac{1}{5} & \cdot \\ -1 & -\frac{1}{3} & \frac{8}{5} & -\frac{5}{8} \\ \cdot & \cdot & -1 & \frac{3}{8} \end{bmatrix}$$

Therefore, the lower and upper triangular factor matrices are:

$$\mathbf{L} = \begin{bmatrix} 3 & \cdot & \cdot & \cdot \\ -1 & \frac{5}{3} & \cdot & \cdot \\ -1 & -\frac{1}{3} & \frac{8}{5} & \cdot \\ \cdot & \cdot & -1 & \frac{3}{8} \end{bmatrix} \quad \mathbf{H} = \begin{bmatrix} 1 & -\frac{1}{3} & -\frac{1}{3} & \cdot \\ \cdot & 1 & -\frac{1}{5} & \cdot \\ \cdot & \cdot & 1 & -\frac{5}{8} \\ \cdot & \cdot & \cdot & 1 \end{bmatrix}$$

The elements of **Y** can be found by solving equation 4.13 using forward substitution. This gives:

$$\begin{bmatrix} 3 & \cdot & \cdot & \cdot \\ -1 & \frac{5}{3} & \cdot & \cdot \\ -1 & -\frac{1}{3} & \frac{8}{5} & \cdot \\ \cdot & \cdot & -1 & \frac{3}{8} \end{bmatrix} \begin{bmatrix} y_1 \\ y_2 \\ y_3 \\ y_4 \end{bmatrix} = \begin{bmatrix} 1 \\ 1 \\ 1 \\ 1 \end{bmatrix}$$

and thus:

$$y_1 = \frac{1}{3}, y_2 = \frac{4}{5}, y_3 = 1 \text{ and } y_4 = \frac{16}{3}$$

The elements of **X** can then be found by solving equations 4.12 using backward substitution. This gives:

$$\begin{bmatrix} 1 & -\frac{1}{3} & -\frac{1}{3} & \cdot \\ \cdot & 1 & -\frac{1}{5} & \cdot \\ \cdot & \cdot & 1 & -\frac{5}{8} \\ \cdot & \cdot & \cdot & 1 \end{bmatrix} \begin{bmatrix} x_1 \\ x_2 \\ x_3 \\ x_4 \end{bmatrix} = \begin{bmatrix} \frac{1}{3} \\ \frac{4}{5} \\ 1 \\ \frac{16}{3} \end{bmatrix}$$

and thus:

$$x_4 = \frac{16}{3}, x_3 = \frac{13}{3}, x_2 = \frac{5}{3} \text{ and } x_1 = \frac{7}{3}$$

4.3.2 LDH factorization

The triangular decomposition technique, described in the preceding section, forms the basis of many modern and efficient techniques. It is evident from the previous numerical example, however, that the elements in the i-th column of **L** are different to those in the i-th row of **H**. This means that both **L** and **H** must be known explicitly and both triangular matrices must be stored. This problem can be alleviated in the case of a symmetrical coefficient matrix **A** by decomposing further the lower triangular matrix **L**. This method, generally known as LDH factorization, expresses the original coefficient matrix **A** as a product of three factor matrices such that:

$$\mathbf{A} = \mathbf{L'DH} \tag{4.17}$$

where **L′** = a lower triangular matrix which has unity elements on its diagonal
H = a higher (or upper) triangular matrix which has unity elements on its diagonal
D = a diagonal matrix which has zero off-diagonal elements.

To achieve this decomposition, the original coefficient matrix **A** is first factorized into the lower and upper triangular matrices **L** and **H**, as described in the preceding section. The matrix **L** is then factorized into **L′** and **D**. The diagonal matrix **D** simply consists of the diagonal elements of **L**. The new lower triangular matrix **L′** is obtained from **L** by dividing each column of **L** by the diagonal element of that column. This technique is best illustrated by considering the example of triangular factorization shown previously (see page 68). This example gave:

$$\mathbf{A} = \begin{bmatrix} 3 & -1 & -1 & \cdot \\ -1 & 2 & \cdot & \cdot \\ -1 & \cdot & 2 & -1 \\ \cdot & \cdot & -1 & 1 \end{bmatrix}, \quad \mathbf{L} = \begin{bmatrix} 3 & \cdot & \cdot & \cdot \\ -1 & \frac{5}{3} & \cdot & \cdot \\ -1 & -\frac{1}{3} & \frac{8}{5} & \cdot \\ \cdot & \cdot & -1 & \frac{3}{8} \end{bmatrix}, \quad \mathbf{H} = \begin{bmatrix} 1 & -\frac{1}{3} & -\frac{1}{3} & \cdot \\ \cdot & 1 & -\frac{1}{5} & \cdot \\ \cdot & \cdot & 1 & -\frac{5}{8} \\ \cdot & \cdot & \cdot & 1 \end{bmatrix}$$

Using the technique described above, $\mathbf{L}'$ and $\mathbf{D}$ can be obtained from $\mathbf{L}$ and are:

$$\mathbf{L}' = \begin{bmatrix} 1 & \cdot & \cdot & \cdot \\ -\frac{1}{3} & 1 & \cdot & \cdot \\ -\frac{1}{3} & -\frac{1}{5} & 1 & \cdot \\ \cdot & \cdot & -\frac{5}{8} & 1 \end{bmatrix} \quad \mathbf{D} = \begin{bmatrix} 3 & \cdot & \cdot & \cdot \\ \cdot & \frac{5}{3} & \cdot & \cdot \\ \cdot & \cdot & \frac{8}{5} & \cdot \\ \cdot & \cdot & \cdot & \frac{3}{8} \end{bmatrix}$$

It is clearly evident from this numerical example that $\mathbf{L}'$ and $\mathbf{H}$ are the transpose of each other provided the original coefficient matrix $\mathbf{A}$ is symmetrical. This is the chief benefit of the decomposition technique, since it is sufficient to obtain and store either the lower triangular matrix $\mathbf{L}'$ or the upper triangular matrix $\mathbf{H}$; the one not stored is known implicitly.

If $\mathbf{A}$ is the coefficient matrix of a set of equations:

$$\mathbf{AX} = \mathbf{b} \tag{4.18}$$

and $\mathbf{A}$ is expressed as:

$$\mathbf{A} = \mathbf{L}'\mathbf{DH} \tag{4.19}$$

then from equations 4.18 and 4.19,

$$\mathbf{L}'\mathbf{DHX} = \mathbf{b} \tag{4.20}$$

Letting $\mathbf{HX} = \mathbf{Y}$ (4.21)

and $\mathbf{DY} = \mathbf{Y}'$ (4.22)

then from equations 4.20–4.22,

$$\mathbf{L}'\mathbf{Y}' = \mathbf{b} \tag{4.23}$$

The elements of $\mathbf{Y}'$ can now be obtained from $\mathbf{L}'$ and $\mathbf{b}$ by forward substitution using equation 4.23, the elements of $\mathbf{Y}$ from $\mathbf{D}$ and $\mathbf{Y}'$ using equation 4.22, which involves only trivial division, and finally the elements of $\mathbf{X}$ can be obtained from $\mathbf{H}$ and $\mathbf{Y}$ by backward substitution using equation 4.21.

4.4 Bi-factorization

The two factorization methods discussed in the preceding sections, i.e. product form of the inverse (section 4.2) and triangular factorization (section 4.3), are both very useful and well established for solving large sets of linear equations, particularly when it is necessary to solve the same basic equations but with different right-hand-side vectors. Although both techniques involve factorization, the method of expressing these factors is considerably different. Zollenkopf combined these two techniques and called it the *bi-factorization* method.

This method is particularly suitable for sparse coefficient matrices that have dominant and non-zero diagonal elements and that are either symmetrical, or, if not symmetrical, have a symmetrical sparsity structure. Although there are very many problems which do not have these features, there are conversely many systems that do. Typical examples are electrical networks, structural analysis, flow systems, etc., and the bi-factorization method is an important and frequently used technique for solving large engineering problems such as these.

The method is based on finding $2n$ factor matrices for an n-th order problem, such that the product of these factor matrices satisfies the requirement:

$$\mathbf{L}^{(n)}\mathbf{L}^{(n-1)} \ldots \mathbf{L}^{(2)}\mathbf{L}^{(1)}\,\mathbf{A}\mathbf{R}^{(1)}\mathbf{R}^{(2)} \ldots \mathbf{R}^{(n-1)}\mathbf{R}^{(n)} = \mathbf{U} \qquad (4.24)$$

where $\mathbf{A}$ = original coefficient matrix
$\mathbf{L}$ = left-hand factor matrices
$\mathbf{R}$ = right-hand factor matrices
$\mathbf{U}$ = unit matrix of order n

Pre-multiplying equation 4.24 by the inverses of $\mathbf{L}^{(n)}$, $\mathbf{L}^{(n-1)} \ldots \mathbf{L}^{(2)}$ and $\mathbf{L}^{(1)}$ consecutively gives:

$$\mathbf{A}\mathbf{R}^{(1)}\mathbf{R}^{(2)} \ldots \mathbf{R}^{(n-1)}\mathbf{R}^{(n)} = (\mathbf{L}^{(1)})^{-1}(\mathbf{L}^{(2)})^{-1} \ldots (\mathbf{L}^{(n-1)})^{-1}(\mathbf{L}^{(n)})^{-1} \qquad (4.25)$$

Post-multiplying equation 4.25 by $\mathbf{L}^{(n)}$, $\mathbf{L}^{(n-1)} \ldots \mathbf{L}^{(2)}$ and $\mathbf{L}^{(1)}$ consecutively gives:

$$\mathbf{A}\mathbf{R}^{(1)}\mathbf{R}^{(2)} \ldots \mathbf{R}^{(n-1)}\mathbf{R}^{(n)}\mathbf{L}^{(n)}\mathbf{L}^{(n-1)} \ldots \mathbf{L}^{(2)}\mathbf{L}^{(1)} = \mathbf{U} \qquad (4.26)$$

Finally, pre-multiplying equation 4.26 by $\mathbf{A}^{-1}$ gives:

$$\mathbf{R}^{(1)}\mathbf{R}^{(2)} \ldots \mathbf{R}^{(n-1)}\mathbf{R}^{(n)}\mathbf{L}^{(n)}\mathbf{L}^{(n-1)} \ldots \mathbf{L}^{(2)}\mathbf{L}^{(1)} = \mathbf{A}^{-1} \qquad (4.27)$$

The factor matrices obtained from the criterion given by equation 4.24, therefore, enable the inverse of the coefficient matrix $\mathbf{A}$ to be expressed and determined implicitly in terms of these factor matrices.

To determine the factor matrices **L** and **R**, the following intermediate matrices are introduced:

$$\mathbf{A} = \mathbf{A}^{(0)}$$

$$\mathbf{A}^{(1)} = \mathbf{L}^{(1)}\mathbf{A}^{(0)}\mathbf{R}^{(1)}$$

$$\mathbf{A}^{(2)} = \mathbf{L}^{(2)}\mathbf{A}^{(1)}\mathbf{R}^{(2)}$$

$$\mathbf{A}^{(k)} = \mathbf{L}^{(k)}\mathbf{A}^{(k-1)}\mathbf{R}^{(k)}$$

$$\mathbf{A}^{(n)} = \mathbf{L}^{(n)}\mathbf{A}^{(n-1)}\mathbf{R}^{(n)}$$

The successive triple inner products $\mathbf{L}^{(k)}\mathbf{A}^{(k-1)}\mathbf{R}^{(k)}$ transform the initial coefficient matrix $\mathbf{A} = \mathbf{A}^{(0)}$ to a unit matrix **U**.

The production of the factor matrices $\mathbf{L}^{(k)}$ and $\mathbf{R}^{(k)}$ and the reduction of the coefficient matrix **A** can be illustrated by considering a fourth-order problem which has an initial coefficient matrix:

$$\mathbf{A} = \mathbf{A}^{(0)} = \begin{bmatrix} a_{11}{}^{(0)} & a_{12}{}^{(0)} & a_{13}{}^{(0)} & a_{14}{}^{(0)} \\ a_{21}{}^{(0)} & a_{22}{}^{(0)} & a_{23}{}^{(0)} & a_{24}{}^{(0)} \\ a_{31}{}^{(0)} & a_{32}{}^{(0)} & a_{33}{}^{(0)} & a_{34}{}^{(0)} \\ a_{41}{}^{(0)} & a_{42}{}^{(0)} & a_{43}{}^{(0)} & a_{44}{}^{(0)} \end{bmatrix}$$

The first reduction step, $\mathbf{L}^{(1)}\mathbf{A}^{(0)}\mathbf{R}^{(1)} = \mathbf{A}^{(1)}$, gives:

$$\begin{bmatrix} L_{11}{}^{(1)} & \cdot & \cdot & \cdot \\ L_{21}{}^{(1)} & 1 & \cdot & \cdot \\ L_{31}{}^{(1)} & \cdot & 1 & \cdot \\ L_{41}{}^{(1)} & \cdot & \cdot & 1 \end{bmatrix} \begin{bmatrix} a_{11}{}^{(0)} & a_{12}{}^{(0)} & a_{13}{}^{(0)} & a_{14}{}^{(0)} \\ a_{21}{}^{(0)} & a_{22}{}^{(0)} & a_{23}{}^{(0)} & a_{24}{}^{(0)} \\ a_{31}{}^{(0)} & a_{32}{}^{(0)} & a_{33}{}^{(0)} & a_{34}{}^{(0)} \\ a_{41}{}^{(0)} & a_{42}{}^{(0)} & a_{43}{}^{(0)} & a_{44}{}^{(0)} \end{bmatrix} \times$$

$$\begin{bmatrix} 1 & R_{12}{}^{(1)} & R_{13}{}^{(1)} & R_{14}{}^{(1)} \\ \cdot & 1 & \cdot & \cdot \\ \cdot & \cdot & 1 & \cdot \\ \cdot & \cdot & \cdot & 1 \end{bmatrix} = \begin{bmatrix} 1 & 0 & 0 & 0 \\ 0 & a_{22}{}^{(1)} & a_{23}{}^{(1)} & a_{24}{}^{(1)} \\ 0 & a_{32}{}^{(1)} & a_{33}{}^{(1)} & a_{34}{}^{(1)} \\ 0 & a_{42}{}^{(1)} & a_{43}{}^{(1)} & a_{44}{}^{(1)} \end{bmatrix}$$

where:

$$L_{11}^{(1)} = \frac{1}{a_{11}^{(0)}}$$

$$L_{i1}^{(1)} = -\frac{a_{i1}^{(0)}}{a_{11}^{(0)}} \qquad R_{1i}^{(1)} = -\frac{a_{1i}^{(0)}}{a_{11}^{(0)}} \qquad (i = 2, 3, 4)$$

$$a_{ij}^{(1)} = a_{ij}^{(0)} - \frac{a_{i1}^{(0)} a_{1j}^{(0)}}{a_{11}^{(0)}} \qquad (i = 2, 3, 4, \quad j = 2, 3, 4)$$

The second reduction step, $\mathbf{L}^{(2)}\, \mathbf{A}^{(1)}\, \mathbf{R}^{(2)} = \mathbf{A}^{(2)}$, gives:

$$\begin{bmatrix} 1 & 0 & \cdot & \cdot \\ \cdot & L_{22}^{(2)} & \cdot & \cdot \\ \cdot & L_{32}^{(2)} & 1 & \cdot \\ \cdot & L_{42}^{(2)} & \cdot & 1 \end{bmatrix} \begin{bmatrix} 1 & \cdot & \cdot & \cdot \\ \cdot & a_{22}^{(1)} & a_{23}^{(1)} & a_{24}^{(1)} \\ \cdot & a_{32}^{(1)} & a_{33}^{(1)} & a_{34}^{(1)} \\ \cdot & a_{42}^{(1)} & a_{43}^{(1)} & a_{44}^{(1)} \end{bmatrix} \times$$

$$\begin{bmatrix} 1 & \cdot & \cdot & \cdot \\ 0 & 1 & R_{23}^{(2)} & R_{33}^{(2)} \\ \cdot & \cdot & 1 & \cdot \\ \cdot & \cdot & \cdot & 1 \end{bmatrix} = \begin{bmatrix} 1 & \cdot & \cdot & \cdot \\ \cdot & 1 & \cdot & \cdot \\ \cdot & \cdot & a_{33}^{(2)} & a_{34}^{(2)} \\ \cdot & \cdot & a_{43}^{(2)} & a_{44}^{(2)} \end{bmatrix}$$

where:

$$L_{22}^{(2)} = \frac{1}{a_{22}^{(1)}}$$

$$L_{i2}^{(2)} = -\frac{a_{i2}^{(1)}}{a_{22}^{(1)}}, \qquad R_{2i}^{(2)} = -\frac{a_{2i}^{(1)}}{a_{22}^{(1)}} \qquad (i = 3, 4)$$

$$a_{ij}^{(2)} = a_{ij}^{(1)} - \frac{a_{i2}^{(1)} a_{2j}^{(1)}}{a_{22}^{(1)}} \qquad (i = 3, 4, \quad j = 3, 4)$$

The fourth and final reduction step, $\mathbf{L}^{(4)}\,\mathbf{A}^{(3)}\,\mathbf{R}^{(4)} = \mathbf{A}^{(4)} = \mathbf{U}$, gives:

$$\begin{bmatrix} 1 & \cdot & \cdot & 0 \\ \cdot & 1 & \cdot & 0 \\ \cdot & \cdot & 1 & 0 \\ \cdot & \cdot & \cdot & L_{44}{}^{(4)} \end{bmatrix} \begin{bmatrix} 1 & \cdot & \cdot & \cdot \\ \cdot & 1 & \cdot & \cdot \\ \cdot & \cdot & 1 & \cdot \\ \cdot & \cdot & \cdot & a_{44}{}^{(3)} \end{bmatrix} \begin{bmatrix} 1 & \cdot & \cdot & \cdot \\ \cdot & 1 & \cdot & \cdot \\ \cdot & \cdot & 1 & \cdot \\ 0 & 0 & 0 & 1 \end{bmatrix} = \begin{bmatrix} 1 & \cdot & \cdot & \cdot \\ \cdot & 1 & \cdot & \cdot \\ \cdot & \cdot & 1 & \cdot \\ \cdot & \cdot & \cdot & 1 \end{bmatrix}$$

where:

$$L_{44}{}^{(4)} = \frac{1}{a_{44}{}^{(3)}}$$

From this fourth-order problem it can be seen that the final **R** matrix is a unit matrix. Furthermore, it can be deduced that, for a more general n-th order problem, the factor matrices **L** and **R** at the k-th reduction step are given by:

$$\mathbf{L}^{(k)} = \begin{bmatrix} 1 & & & 0 & & \\ & \ddots & & \vdots & & \\ & & 1 & 0 & & \\ & & & \mathrm{L}_{kk}{}^{(k)} & & \\ & & & \mathrm{L}_{ik}{}^{(k)} & 1 & \\ & & & \vdots & & \ddots \\ & & & \mathrm{L}_{nk}{}^{(k)} & & & 1 \end{bmatrix}$$

$$\mathbf{R}^{(k)} = \begin{bmatrix} 1 & & & & & & \\ & \ddots & & & & & \\ & & 1 & & & & \\ 0 & \cdots & 0 & 1 & \mathrm{R}_{kj}{}^{(k)} & \cdots & \mathrm{R}_{kn}{}^{(k)} \\ & & & & 1 & & \\ & & & & & \ddots & \\ & & & & & & 1 \end{bmatrix}$$

where:

$$L_{kk}{}^{(k)} = \frac{1}{a_{kk}{}^{(k-1)}}$$

$$L_{ik}^{(k)} = -\frac{a_{ik}^{(k-1)}}{a_{kk}^{(k-1)}} \qquad (i = k+1, \ldots, n)$$

$$R_{kj}^{(k)} = -\frac{a_{kj}^{(k-1)}}{a_{kk}^{(k-1)}} \qquad (j = k+1, \ldots, n)$$

$$a_{ij}^{(k)} = a_{ij}^{(k-1)} - \frac{a_{ik}^{(k-1)} a_{kj}^{(k-1)}}{a_{kk}^{(k-1)}} \qquad \begin{matrix}(i = k+1, \ldots, n)\\(j = k+1, \ldots, n)\end{matrix}$$

For a symmetrical matrix **A**:

$$a_{ik}^{(k-1)} = a_{ki}^{(k-1)}$$

Therefore,

$$R_{ik}^{(k)} = L_{ki}^{(k)} \tag{4.28}$$

In the case of symmetrical coefficient matrices, equation 4.28 indicates that, except for the diagonal elements, the k-th row of $\mathbf{R}^{(k)}$ is identical to the k-th column of $\mathbf{L}^{(k)}$. Also, the diagonal elements of $\mathbf{R}^{(k)}$ are all unity, and, since these are known implicitly, it is sufficient only to evaluate the elements of $\mathbf{L}^{(k)}$. Therefore, the required number of operations and the amount of storage space is reduced to almost a half.

To illustrate the bi-factorization technique, consider the numerical solution to the previous example which was:

$$\begin{bmatrix} 3 & -1 & -1 & \cdot \\ -1 & 2 & \cdot & \cdot \\ -1 & \cdot & 2 & -1 \\ \cdot & \cdot & -1 & 1 \end{bmatrix} \begin{bmatrix} x_1 \\ x_2 \\ x_3 \\ x_4 \end{bmatrix} = \begin{bmatrix} 1 \\ 1 \\ 1 \\ 1 \end{bmatrix}$$

The first reduction step, $\mathbf{L}^{(1)}\,\mathbf{A}^{(0)}\,\mathbf{R}^{(1)} = \mathbf{A}^{(1)}$, gives:

$$\begin{bmatrix} \frac{1}{3} & \cdot & \cdot & \cdot \\ \frac{1}{3} & 1 & \cdot & \cdot \\ \frac{1}{3} & \cdot & 1 & \cdot \\ 0 & \cdot & \cdot & 1 \end{bmatrix} \begin{bmatrix} 3 & -1 & -1 & \cdot \\ -1 & 2 & \cdot & \cdot \\ -1 & \cdot & 2 & -1 \\ \cdot & \cdot & -1 & 1 \end{bmatrix} \begin{bmatrix} 1 & \frac{1}{3} & \frac{1}{3} & 0 \\ \cdot & 1 & \cdot & \cdot \\ \cdot & \cdot & 1 & \cdot \\ \cdot & \cdot & \cdot & 1 \end{bmatrix} = \begin{bmatrix} 1 & \cdot & \cdot & \cdot \\ \cdot & \frac{5}{3} & -\frac{1}{3} & \cdot \\ \cdot & -\frac{1}{3} & \frac{5}{3} & -1 \\ \cdot & \cdot & -1 & 1 \end{bmatrix}$$

The second reduction step, $\mathbf{L}^{(2)}\,\mathbf{A}^{(1)}\,\mathbf{R}^{(2)} = \mathbf{A}^{(2)}$, gives:

$$\begin{bmatrix} 1 & 0 & \cdot & \cdot \\ \cdot & \frac{3}{5} & \cdot & \cdot \\ \cdot & \frac{1}{5} & 1 & \cdot \\ \cdot & 0 & \cdot & 1 \end{bmatrix} \begin{bmatrix} 1 & \cdot & \cdot & \cdot \\ \cdot & \frac{5}{3} & -\frac{1}{3} & \cdot \\ \cdot & -\frac{1}{3} & \frac{5}{3} & -1 \\ \cdot & \cdot & -1 & 1 \end{bmatrix} \begin{bmatrix} 1 & \cdot & \cdot & \cdot \\ 0 & 1 & \frac{1}{5} & 0 \\ \cdot & \cdot & 1 & \cdot \\ \cdot & \cdot & \cdot & 1 \end{bmatrix} = \begin{bmatrix} 1 & \cdot & \cdot & \cdot \\ \cdot & 1 & \cdot & \cdot \\ \cdot & \cdot & \frac{8}{5} & -1 \\ \cdot & \cdot & -1 & 1 \end{bmatrix}$$

The third reduction step, $\mathbf{L}^{(3)}\,\mathbf{A}^{(2)}\,\mathbf{R}^{(3)} = \mathbf{A}^{(3)}$, gives:

$$\begin{bmatrix} 1 & \cdot & 0 & \cdot \\ \cdot & 1 & 0 & \cdot \\ \cdot & \cdot & \frac{5}{8} & \cdot \\ \cdot & \cdot & \frac{5}{8} & 1 \end{bmatrix} \begin{bmatrix} 1 & \cdot & \cdot & \cdot \\ \cdot & 1 & \cdot & \cdot \\ \cdot & \cdot & \frac{8}{5} & -1 \\ \cdot & \cdot & -1 & 1 \end{bmatrix} \begin{bmatrix} 1 & \cdot & \cdot & \cdot \\ \cdot & 1 & \cdot & \cdot \\ 0 & 0 & 1 & \frac{5}{8} \\ \cdot & \cdot & \cdot & 1 \end{bmatrix} = \begin{bmatrix} 1 & \cdot & \cdot & \cdot \\ \cdot & 1 & \cdot & \cdot \\ \cdot & \cdot & 1 & \cdot \\ \cdot & \cdot & \cdot & \frac{3}{8} \end{bmatrix}$$

The fourth and final reduction step, $\mathbf{L}^{(4)}\,\mathbf{A}^{(3)}\,\mathbf{R}^{(4)} = \mathbf{A}^{(4)} = \mathbf{U}$, gives:

$$\begin{bmatrix} 1 & \cdot & \cdot & 0 \\ \cdot & 1 & \cdot & 0 \\ \cdot & \cdot & 1 & 0 \\ \cdot & \cdot & \cdot & \frac{8}{3} \end{bmatrix} \begin{bmatrix} 1 & \cdot & \cdot & \cdot \\ \cdot & 1 & \cdot & \cdot \\ \cdot & \cdot & 1 & \cdot \\ \cdot & \cdot & \cdot & \frac{3}{8} \end{bmatrix} \begin{bmatrix} 1 & \cdot & \cdot & \cdot \\ \cdot & 1 & \cdot & \cdot \\ \cdot & \cdot & 1 & \cdot \\ 0 & 0 & 0 & 1 \end{bmatrix} = \begin{bmatrix} 1 & \cdot & \cdot & \cdot \\ \cdot & 1 & \cdot & \cdot \\ \cdot & \cdot & 1 & \cdot \\ \cdot & \cdot & \cdot & 1 \end{bmatrix}$$

The solution to the original equations can now be found since, from equation 4.27,

$$\mathbf{A}^{-1} = \mathbf{R}^{(1)}\,\mathbf{R}^{(2)}\,\mathbf{R}^{(3)}\,\mathbf{R}^{(4)}\,\mathbf{L}^{(4)}\,\mathbf{L}^{(3)}\,\mathbf{L}^{(2)}\,\mathbf{L}^{(1)}$$

and, since $\mathbf{R}^{(4)}$ is a unit matrix:

$$\mathbf{X} = \mathbf{R}^{(1)}\,\mathbf{R}^{(2)}\,\mathbf{R}^{(3)}\,\mathbf{L}^{(4)}\,\mathbf{L}^{(3)}\,\mathbf{L}^{(2)}\,\mathbf{L}^{(1)}\,\mathbf{b}$$

Multiplication of these factor matrices gives:

$$\mathbf{X} = \begin{bmatrix} \frac{7}{3} \\ \\ \frac{5}{3} \\ \\ \frac{13}{3} \\ \\ \frac{16}{3} \end{bmatrix}$$

This numerical example shows that, even for a small fourth-order problem, the factor matrices are sparse. This indicates that for large problems with a high degree of sparsity, for which this technique is ideally suited, the gains are very significant.

To calculate **X**, the **b** vector has to be multiplied sequentially by $2n$ factor matrices. However, except for either one row or one column, these factors are all unit matrices and therefore each of the individual multiplications becomes quite trivial and with suitable programming involves only the non-zero elements of each factor matrix.

4.5 Comparison between triangulation and bi-factorization methods

The numerical processes involved in the triangular decomposition and bi-factorization methods are basically identical, being dependent on the Gauss elimination technique. Therefore, the total number of arithmetical operations required to factorize the coefficient matrix of a given set of simultaneous linear equations and for a given sequence of elimination are the same with both methods. On this basis alone neither method has any advantage over the other. When a digital computer is used for numerically solving the equations, however, the bi-factorization method does achieve a basic advantage over triangular decomposition. This is because the factors created by the two methods are different; this difference being such that the bi-factorization method can require less storage, less indexing and simpler arithmetic operations to solve the equations from a knowledge of the factors.

To illustrate some of these differences, consider the factor matrices produced in the previous numerical examples. The factor matrices for both methods can be superimposed if the unity diagonals of the upper triangular matrix and the right-hand factor matrix are ignored; these being known implicitly. The resulting arrays are as shown in Fig. 4.1.

From Fig. 4.1 it is evident that, although the original coefficient matrix of the equations was symmetrical, the superimposed factor matrices of triangular decomposition are asymmetrical whereas those of the bi-factorization are

$$\begin{bmatrix} 3 & -\frac{1}{3} & -\frac{1}{3} & \cdot \\ -1 & \frac{5}{3} & -\frac{1}{5} & \cdot \\ -1 & -\frac{1}{3} & \frac{8}{5} & -\frac{5}{8} \\ \cdot & \cdot & -1 & \frac{3}{8} \end{bmatrix} \qquad \begin{bmatrix} \frac{1}{3} & \frac{1}{3} & \frac{1}{3} & \cdot \\ \frac{1}{3} & \frac{3}{5} & \frac{1}{5} & \cdot \\ \frac{1}{3} & \frac{1}{5} & \frac{5}{8} & \frac{5}{8} \\ \cdot & \cdot & \frac{5}{8} & \frac{8}{3} \end{bmatrix}$$

(a) (b)

Fig. 4.1 Factor matrices (*a*) triangular decomposition; (*b*) bi-factorization

symmetrical. In the case of bi-factorization, therefore, only the left-hand factor matrices need be stored, the right-hand ones being known implicitly. In the case of triangular decomposition, both the lower and upper triangular matrices must be stored, which nearly doubles the required storage space and indexing information.

This inherent problem of triangular decomposition can be overcome by further decomposing the LH form of triangulation into the LDH form. The storage is now reduced to that of bi-factorization but, to achieve this, an additional set of operations is required, which, itself, decreases computational efficiency.

The other difficulty with triangular decomposition is that the product of its two factor matrices gives the original coefficient matrix (see section 4.9) and not its inverse. On the other hand, the product of the factor matrices produced by bi-factorization does give the inverse (see equation 4.27). Consequently, to solve the equations for the unknown quantities after factorization is more difficult using triangular decomposition than using bi-factorization.

Despite these various differences, both methods can be made very efficient and much superior to direct matrix inversion. To be made efficient, however, both methods require skilful programming and the inclusion of, for instance, sparsity techniques. Furthermore, the relative differences between the two methods can be reversed quite significantly, depending upon the quality and computational efficiency of the programming, since this is only marginally related to the mathematical differences between the two methods.

Bibliography

Berezin, O. M. & Zhidkov, N. P.: *Computing Methods,* vol. II, chap. 6. Pergamon Press, 1965.

Brameller, A. & Allan, R. N.: The role of sparsity in the analysis of large systems. *Comput. Aided Des.* **6**, 159–68, 1974.

Dantzig, G. B.: *Linear Programming and Extensions.* Princeton University Press, 1963.

Dantzig, G. B. & Orchard-Hays, W.: The product form of inverse in the simplex method. *Math. Computn* **8**, 64–7, 1954.

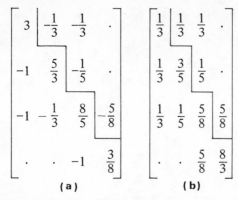

Forsythe, G. E. & Moler, C. B.: *Computer Solution of Linear Algebraic Equations.* Prentice-Hall, 1967.
Fox, L: *Introduction to Numerical Linear Algebra.* Clarendon Press, 1965.
Gass, S.: *Linear Programming, Methods and Applications.* McGraw-Hill, 1958.
Hadley, G.: *Linear Programming.* Addison-Wesley, 1962.
Ralston, A. & Wilf, H. S.: *Mathematical Methods for Digital Computers.* John Wiley, 1960.
Tewarson, R. P.: On the product form of inverses of sparse matrices. *Symp. appl. Math. Rev.* **8**, 336–42, 1966.
Tewarson, R. P.: *Sparse Matrices.* Academic Press, 1973.
Tinney, W. F. & Walker, J. W.: Direct solutions of sparse network equations by optimally ordered triangular factorisation. *Proc. Inst. elect. electron. Engrs* **55**, 1801–9, 1967.
Zollenkopf, K.: Bi-factorisation: basic computational algorithm and programming techniques. In *Large Sparse Sets of Linear Equations,* pp. 75–96. Academic Press, 1971.

5
Sparsity-directed Elimination

5.1 Introduction

The preceding chapters of this book showed how the elimination techniques could be applied to solve directly equations of the form **AX** = **b**. They also showed that if the coefficient matrix **A** is sparse, the various factored forms of this matrix can possess a high degree of sparsity, that is, the relative number of non-zero elements in these matrices could be small. It is clear that the efficiency of the solution would be increased if only these non-zero elements are stored and processed. In these preceding chapters, however, no attempt was made to discuss in detail how this property of sparsity could be exploited nor how the elimination process should be used to maintain this degree of sparsity.

An unfortunate property of the elimination or factorization process of sparse matrices is that new non-zero elements can be continuously generated. This would decrease the original degree of sparsity and is known as fill-in. It is clearly desirable to keep the number of new elements to a minimum in order to minimize the required storage space and the total computation time.

The amount of fill-in depends greatly on the order in which the pivotal rows and columns are selected. There are a number of available techniques which can be used to select a suitable order of elimination and hence reduce the amount of fill-in. These techniques can be very efficient computationally and can produce an order of elimination which is sufficiently close to the optimum to be of considerable practical value in analysing large, sparse networks. The purpose of this chapter is to describe why the new non-zero elements are generated and how these various ordering techniques reduce the amount of fill-in.

5.2 Matrix elimination and graph reduction

As discussed in Chapter 2, the structure of a physical network or system can be represented by a graph, and the behaviour of the system by a set of simultaneous equations. The associated coefficient matrix of these equations will generally be symmetrical but, in certain circumstances, may also be asymmetrical. In both cases, the graph representation can still be made. The set of simultaneous equations may, however, not relate to a physical network or system. In such cases the associated coefficient matrix can still be represented by a graph and therefore network-type structure, even though this graph does not represent a physical system.

Every node in this graph corresponds to a row and column of the matrix from which it was derived. If the matrix is the nodal coefficient matrix of a network, then the corresponding graph is the same as the graph of the original network. If the matrix is the loop coefficient matrix of a network, then its graph is the dual of the original network.

The ability to represent a matrix by a graph can help significantly in two respects. Firstly it assists in recognizing the sparsity structure of the problem, which may be very difficult from the array of numbers in a matrix. Secondly it enables a clear understanding of fill-in to be appreciated and how this unfortunate effect can be minimized during the elimination process.

In a connected graph, for example as shown in Fig. 5.1(*a*), information is passed between the nodes through the interconnecting branches. For information to pass between nodes i and k of Fig. 5.1(*a*), this information has to pass through an intermediate node j, as shown by the solid lines. In this case the elimination of node j would destroy the path between nodes i and k and, for continued transmission of information, a new direct path would have to be introduced as shown by the dotted line. On the other hand, if direct paths exist between all nodes adjacent to node j before elimination, as shown in Fig. 5.1(*b*), node j could be eliminated without affecting the remaining graph. To illustrate these points consider the nodal network shown in Fig. 5.2(*a*). In this network, node *0* is taken as the reference node or datum. The corresponding coefficient matrix **A** and the graph is also shown in Fig. 5.2.

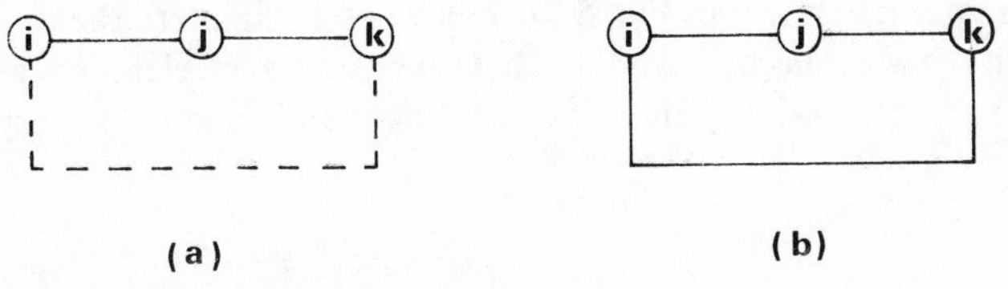

Fig. 5.1 Two typical connected graphs
(*a*) nodes i and k not directly connected; (*b*) nodes i and k directly connected

To illustrate the above points consider the effect of eliminating column 1 and row 1 of the coefficient matrix, that is, eliminate node 1 of the graph. From Fig. 5.2(*c*), it is seen that nodes 2 and 4 are immediate neighbours of node 1 and that there is no direct path between nodes 2 and 4. Therefore, if node 1 is eliminated, a new direct path between nodes 2 and 4 must be created. Consider the partial matrix which contains only column 1 and row 1 and its corresponding graph as shown in Fig. 5.3.

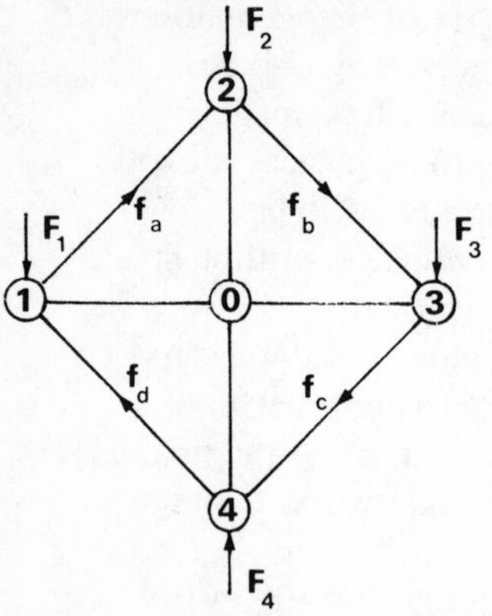

(a)

$$A = \begin{bmatrix} a_{11} & a_{12} & \cdot & a_{14} \\ a_{21} & a_{22} & a_{23} & \cdot \\ \cdot & a_{32} & a_{33} & a_{34} \\ a_{41} & \cdot & a_{43} & a_{44} \end{bmatrix}$$

(b)

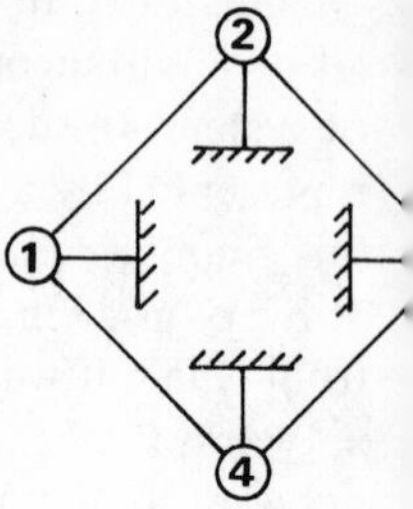

(c)

Fig. 5.2 Nodal network and its graph
(*a*) network; (*b*) coefficient matrix; (*c*) graph
(For clarity, the direction of the graph is not shown and the datum is indicated by the shaded lines.)

$$\left[\begin{array}{c|ccc} X & X & \cdot & X \\ \hline X & \cdot & \cdot & \cdot \\ \cdot & \cdot & \cdot & \cdot \\ X & \cdot & \cdot & \cdot \end{array}\right]$$

(a)

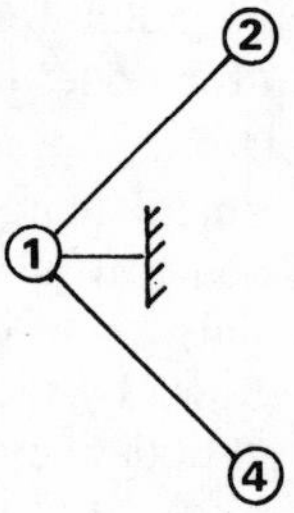

(b)

Fig. 5.3 Graph of partial matrix
(*a*) partial matrix; (*b*) graph

When node 1 of Fig. 5.3(*b*) is eliminated, four new non-zero elements must be added to the matrix shown in Fig. 5.3(*a*) in order to simulate the connection of a direct path between nodes 2 and 4. These new non-zero elements are indicated by $\textcircled{X}$ in Fig. 5.4, which shows the new matrix and its corresponding graph.

$$\left[\begin{array}{c|ccc} X & X & \cdot & X \\ \hline X & \textcircled{X} & \cdot & \textcircled{X} \\ \cdot & \cdot & \cdot & \cdot \\ X & \textcircled{X} & \cdot & \textcircled{X} \end{array}\right]$$

(a)

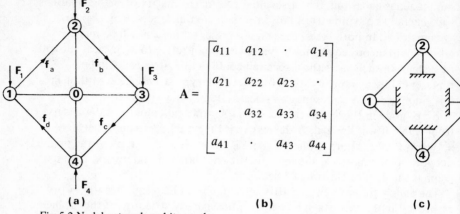

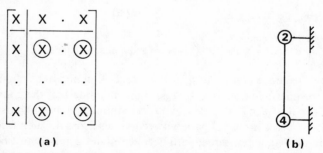

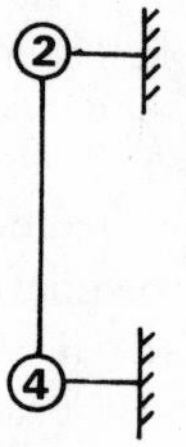

(b)

Fig. 5.4 Introduction of new elements during elimination
(*a*) new partial matrix; (*b*) graph

The elimination of node 1, that is, column 1 and row 1 of the original 4 x 4 matrix, reduces the matrix to a 3 x 3. This reduced matrix is shown in Fig. 5.5 together with the corresponding graph. This new graph can be obtained by deleting the graph shown in Fig. 5.3(*b*) from that shown in Fig. 5.2(*c*) and adding the graph shown in Fig. 5.4(*b*).

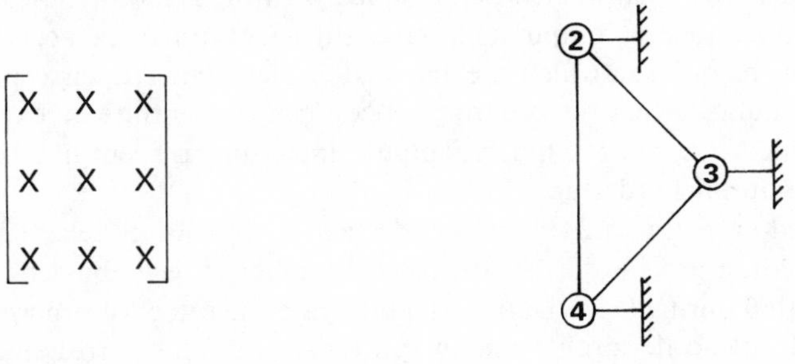

(a) (b)

Fig. 5.5 Effect of eliminating node 1 from Fig. 5.2
(*a*) reduced matrix; (*b*) graph

The diagonal elements of networks are, in general, non-zero. Therefore the elimination process does not create fill-in on the diagonal, and the connections between independent nodes and the datum in the corresponding graph need not be considered. In the example discussed above, only two new non-zero elements would therefore be created in the reduced matrix when node 1 is eliminated, that is, one new path.

The number of new paths introduced by eliminating any given node or group of nodes is defined as the *valency* of that node or group of nodes. Also, during the elimination process, parallel paths in a graph can be represented by an equivalent single path and therefore can be treated as one path only.

5.3 Principles of ordering

There are three main objectives for pivotal ordering in any of the different versions of Gauss elimination. The first is to maintain a high numerical accuracy, the second is to preserve the sparsity of the matrix and the third is to increase computational efficiency. In general these objectives are not necessarily compatible. To maintain maximum numerical accuracy, the highest valued element is chosen as the first pivot; this process is then repeated sequentially. To preserve sparsity, however, this preferable pivotal order may be unsatisfactory and may lead to significant fill-in. Also, to preserve maximum sparsity and therefore minimize storage and the computation time required for Gauss elimination may necessitate excessive computational time for the ordering process. To maintain maximum sparsity may therefore reduce the computational efficiency and decrease numerical accuracy.

In most problems a compromise must be achieved between the various objectives discussed above. In general, however, the physical nature of network problems is such that the diagonal elements of the matrix are usually numerically dominant. Therefore it is usually unnecessary to examine off-diagonal elements and the ordering process can be concentrated on the diagonal elements only. Also, the variation in the values of the diagonal elements is normally sufficiently small that, when coupled with the great numerical accuracy of large modern computers, it gives sufficient numerical accuracy irrespective of the order in which the diagonal elements are chosen as pivots. These aspects indicate that the ordering process can concentrate on exploiting sparsity provided that not too much computational time is spent in attempting to achieve optimal ordering.

As discussed in section 5.2, the order in which the pivots are selected during the elimination process greatly affects the number of new non-zero elements created during factorization. To illustrate the effect of ordering, consider the fourth-order problem shown in Fig. 5.6 and the corresponding graph.

$$\begin{bmatrix} X & X & X & X \\ X & X & \cdot & \cdot \\ X & \cdot & X & \cdot \\ X & \cdot & \cdot & X \end{bmatrix}$$

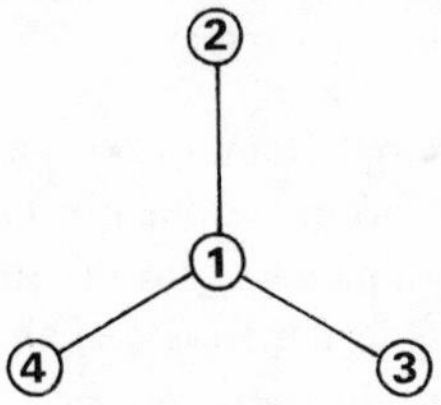

Fig. 5.6 Fourth-order radial network problem

If node 1 is eliminated first, then the reduced matrix and its corresponding graph are as shown in Fig. 5.7.

$$\begin{bmatrix} X & X & X \\ X & X & X \\ X & X & X \end{bmatrix}$$

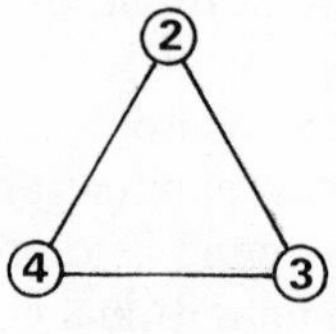

Fig. 5.7 Effect of eliminating node 1 from Fig. 5.6.

It can be seen that this reduced matrix is completely full and the original degree of sparsity has been lost. Further reduction steps cannot recover this original degree of sparsity. This problem has been created because, as discussed in section 5.2, elimination of node 1 necessitates the creation of direct paths between all the other nodes which have been disconnected.

If the nodes of the original graph shown in Fig. 5.6 are re-numbered as shown in Fig. 5.8, then the original degree of sparsity is maintained when node 1 is eliminated, since no new direct paths are required. This effect is shown by the reduced matrix and corresponding graph in Fig. 5.9.

$$\begin{bmatrix} X & \cdot & \cdot & X \\ \cdot & X & \cdot & X \\ \cdot & \cdot & X & X \\ X & X & X & X \end{bmatrix}$$

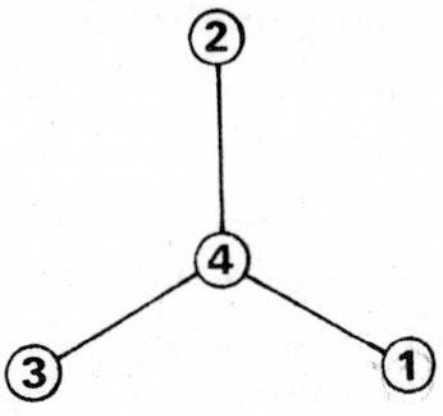

Fig. 5.8 Re-numbered radial network problem

$$\begin{bmatrix} X & \cdot & X \\ \cdot & X & X \\ X & X & X \end{bmatrix}$$

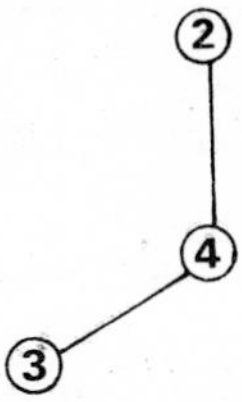

Fig. 5.9 Effect of eliminating node 1 from Fig. 5.8

The above example of a simple radial system illustrates clearly the effect of node numbering on the creation of non-zero elements during elimination, assuming that the nodes are eliminated in natural order. In practice, re-numbering is unnecessary because the nodes can be eliminated in any numerical order. For example, the original degree of sparsity would have been preserved if node 2 of Fig. 5.6 had been eliminated first, followed by nodes 4, 3 and 1. It is also interesting to observe that eleven other orders of elimination could have been used whilst preserving the sparsity of the original problem, the only requirement being that node 1 must be eliminated either last or last but one.

In the above simple example, the preferred order of elimination to maintain sparsity can be assessed by inspection. For large, sparse matrices, this technique is impossible and other techniques have to be used. Several practical methods have been suggested in the literature and some of the more important ones are discussed later in this chapter. It is worth noting, however, that most of these methods are based on the principles discussed in this section.

The simplest method of ordering is to use the natural order of the rows and columns, that is, row one and column one are eliminated first, row two and column two are eliminated second and the process continued until all rows and columns have been eliminated. This scheme has been used frequently in the past. The method, however, is likely to be very non-optimal, since it depends entirely on the random numbering of the nodes in the original equivalent graph. This scheme therefore leads to long factorization times and the creation of a large number of new non-zero elements.

At the other extreme is optimal ordering. This method must deduce the pivotal order which creates the minimum number of new non-zero elements during the complete elimination process, assuming that sufficient numerical accuracy can be preserved. Techniques to achieve this optimal order are idealistic and not very practical because the number of operations which are necessary, and consequently the computational time, becomes very large. Therefore, the gains made in the factorization process are generally outweighed by the losses incurred in selecting the optimal order.

Several practical schemes exist which are compromises between the two extremes discussed above. These schemes are usually referred to as *semi-optimal* methods because, in general, they create an order which is sufficiently close to the optimal order. They can be divided into two categories: *pre-ordering* schemes, in which the order is established before the elimination process is started, and *dynamic* ordering schemes, in which the order is established during the elimination process.

5.4 Pre-ordering techniques

5.4.1 Least number of connected branches

In this method of ordering, the sequence of elimination can be determined by inspecting the original graph of the problem; the sequence being in the order of increasing number of branches connected to each node. If more than one node has the same number of connected branches, then these nodes can be eliminated in any order. To illustrate this method, consider the sparse matrix represented by the graph shown in Fig. 5.10.

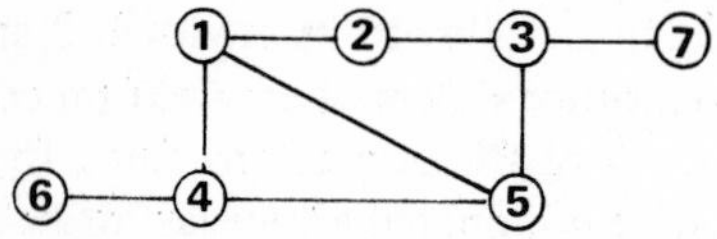

Fig. 5.10 Typical graph of a sparse matrix

In this example, nodes 6 and 7 have only one branch connected to them and therefore, using this method, they must be eliminated first, in any order. Node 2 has two branches connected to it and therefore must be eliminated next. Finally, nodes 1, 3, 4 and 5 all have three branches connected to them and therefore must be eliminated last, in any order. Although for this example several sequences exist, one possible order of elimination is:

(6,7,2,1,3,4,5)

In practice, the number of branches connected to each node need not be determined by inspection of the graph, because this number depends on the number of off-diagonal elements in the corresponding row or column of the associated coefficient matrix. Therefore the order of elimination can be determined by inspecting the number of off-diagonal elements and arranging these in numerical ascending order.

This scheme is very simple to program for a digital computer and the number of operations required to determine the order of elimination is small. The scheme produces the optimal order for a star graph and is usually adequate for the practical solution of small systems. For large problems involving the analysis of hundreds of variables and more, the method becomes unattractive because the number of new non-zero elements increases substantially, and different methods should be used. It may, however, be a reasonable compromise for an investigator studying a one-off problem, rather than spending a considerable time writing and de-bugging the problem for a more complex method. As this situation does not often occur, the method should be viewed as a simple method for simple problems.

5.4.2 Diagonal banding

The purpose of the banding schemes is to arrange the original non-zero elements of a matrix in a narrow band around either the major diagonal or minor diagonal, as shown in Fig. 5.11

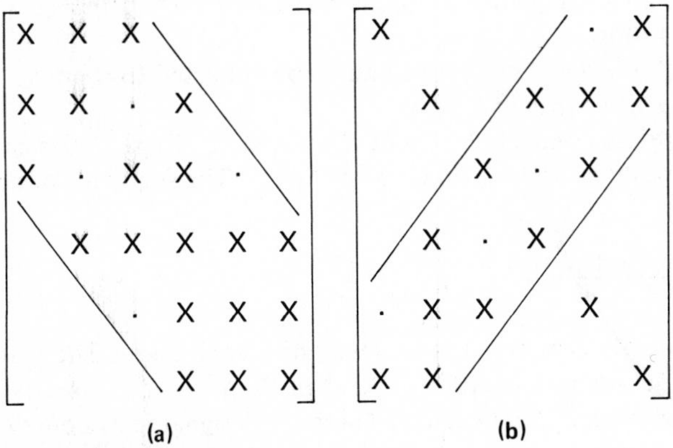

Fig. 5.11 Diagonal banding
(*a*) major diagonal; (*b*) minor diagonal

One of the basic properties of factorization is that no new non-zero elements can be created outside a major diagonal band during the elimination process. Similarly, no new non-zero elements can be created outside a minor diagonal band for special types of problems. Therefore, only the elements within the band need be stored and all arithmetical operations are confined to these elements. However, even zero elements within the band must be stored, since, during the elimination process, it is very likely that these elements will become non-zero. For this reason, the method does not exploit sparsity to the full unless the width of the band, known as *bandwidth,* is relatively narrow and the original number of zero elements in the band is very small.

The logical steps which are necessary in order to arrange the non-zero

elements into a major diagonal band are as follows:
(i) select the node with the least number of branches connected to it and designate it 1. If more than one node conforms to this requirement, any of them can be selected;
(ii) assign the next sequence of node numbers to those nodes directly connected to node 1;
(iii) continue sequentially the process described in (ii) until all the nodes in the system have been designated.

To illustrate the diagonal banding process consider the graph shown in Fig. 5.12.

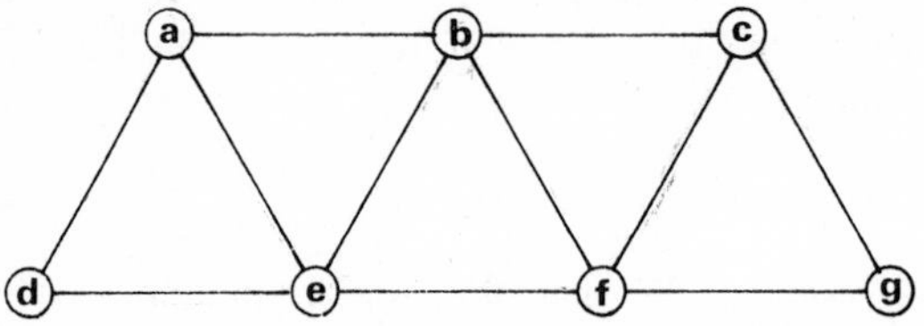

Fig. 5.12 Typical ladder-type network

Following the above sequence of logical steps, nodes *a* to *g* in Fig. 5.12 can be numbered as follows:
(i) the nodes with the least number of branches connected to them are *d* and *g*. Choose node *d* arbitrarily and designate it 1;
(ii) the nodes directly connected to node *d* are *a* and *e*. These nodes can be designated arbitrarily as nodes 2 and 3 respectively. The sequence of nodes so far is therefore:

re-numbered node	1,	2,	3
original node	*d*,	*a*,	*e*

(iii) the nodes directly connected to nodes *a* and *e* are *b* and *f*. These are designated as nodes 4 and 5 in either order. Finally, the nodes directly connected to nodes *b* and *f* are *c* and *g*. These are designated as nodes 6 and 7.

Therefore one possible order for the nodes is

re-numbered node	1,	2,	3,	4,	5,	6,	7
original node	*d*,	*a*,	*e*,	*b*,	*f*,	*c*,	*g*

and the matrix created by this sequence of ordering is shown in Fig. 5.13.

The alternative banding scheme is to arrange the non-zero elements into a band about the minor diagonal. The procedure for achieving this is as follows:
(i) select a node with the least number of branches connected to it and designate it 1;
(ii) select the nodes directly connected to node 1 and assign to them the highest node numbers available, that is, if there are n nodes, these adjacent nodes are designated n, $n-1$, . . .
(iii) select the nodes directly connected to those designated in (ii) and assign

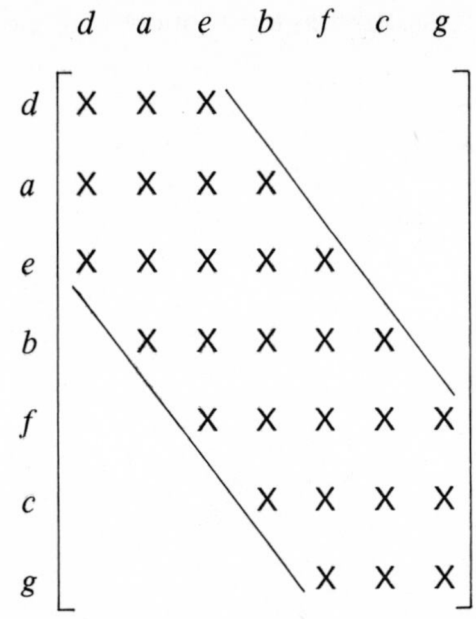

Fig. 5.13 Major diagonal banding for network of Fig. 5.12

to them the next lowest node numbers available, that is, at the third step these adjacent nodes are designated 2, 3, . . .
(iv) continue sequentially the alternating process described in (ii) and (iii) until all the nodes have been designated.

To illustrate this process, consider the graph shown in Fig. 5.14.

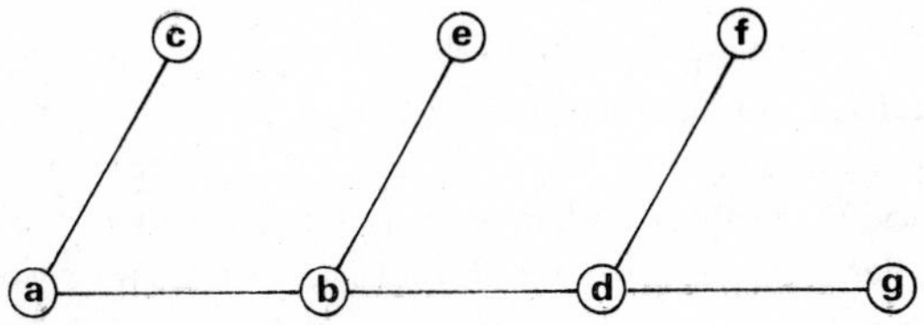

Fig. 5.14 Typical radial-type network

Following the above sequence of logical steps, nodes *a* to *g* in Fig. 5.14 can be numbered as follows:
(i) there are several nodes with only one branch connected to them and any of these can be selected as the first node. Choose arbitrarily node *c* and designate it 1;
(ii) the only node directly connected to node *c* is *a*. The total number of nodes is seven and therefore, at this stage, the highest node number available is 7. Hence node *a* is designated 7;
(iii) the only node directly connected to node *a* is *b*. The next available lowest number is 2 and therefore this is assigned to node *b*;
(iv) the nodes directly connected to node *b* are *d* and *e*. The next available highest node numbers are 5 and 6 and these are assigned in any order to nodes *d* and *e*. Finally, the nodes directly connected to node *d* are *f* and *g*. The next available lowest node numbers are 3 and 4 and these are assigned to nodes *f* and *g*; the order again being arbitrary.

Therefore, although several alternative sequences exist, one possible order is:

re-numbered node	1,	2,	3,	4,	5,	6,	7
original node	*c,*	*b,*	*f,*	*g,*	*d,*	*e,*	*a*

which creates a matrix as shown in Fig. 5.15.

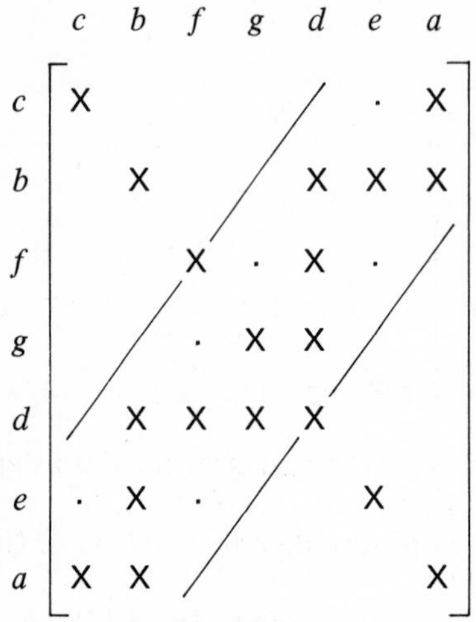

Fig. 5.15 Minor diagonal banding for network of Fig. 5.14

Both the major and minor diagonal banding schemes are efficient only for graphs having special structures. Major diagonal banding is most suitable for ladder-type networks, shown typically in Fig. 5.12, that is, for matrices whose graphs consist of nodes mainly connected in a chain. From Fig. 5.13, it is seen that no zero elements exist in the band, and therefore, since non-zero elements cannot be created outside the band, major diagonal banding has achieved minimum storage and maximum exploitation of sparsity for this particular problem.

With radial-type networks, as shown typically in Fig. 5.14, major diagonal banding allows many zero elements to be stored in the band, which requires unnecessary storage and increased computation time. For instance, if major diagonal banding had been used for the graph shown in Fig. 5.14, one possible matrix would be as shown in Fig. 5.16.

From Fig. 5.16, it is seen that the number of non-zero elements in this banding arrangement is 19 and the number of zero elements is 10, this being a significant proportion of the total. For this type of problem it is more efficient to use minor diagonal banding. For instance, from Fig. 5.15, it is seen that the number of zero elements which must be stored for the same problem is reduced to 6.

For the large, more general type of problem, such as an electrical- or gas-distribution system, both of the banding schemes are likely to have a very sparse structure within the band. Considerable fill-up will therefore take

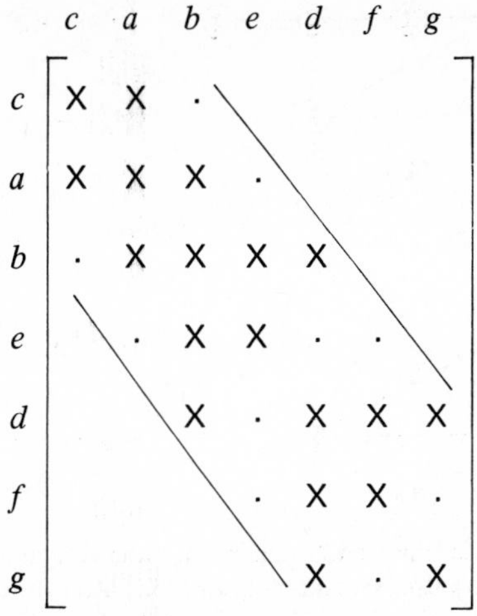

Fig. 5.16 Major diagonal banding for network of Fig. 5.14

place during the elimination process. This significant disadvantage can be reduced by optimizing the bandwidth, but this is at the expense of longer computational times. A much more effective and efficient method to use for these more general large and sparse networks is dynamic ordering.

5.5 Dynamic ordering techniques

The ordering schemes described in section 5.4 established the order of elimination before the reduction process is commenced. During this elimination process, the graph corresponding to the reduced matrix is continuously changing. The pre-ordering schemes discussed previously cannot account for these changes and therefore a reasonably efficient or semi-optimal scheme cannot be guaranteed. To overcome this problem, the corresponding graph obtained at each reduction step should be examined in order to conserve the sparsity of the coefficient matrix at the next reduction step. The schemes achieving this object are known as dynamic ordering schemes. There are several possible methods which have been proposed and tried and the two which are most appropriate for general large, sparse networks are discussed in the following sub-sections.

5.5.1 Least number of connected branches

This technique is the simplest dynamic ordering scheme and the most commonly used. It is identical with the first pre-ordering scheme discussed in section 5.4, except that the technique is applied not only initially but also at each of the reduction steps. Therefore, by extending the previous discussion, this ordering technique is equivalent to the elimination of a node in the reduced graph having the least number of connected branches. This scheme

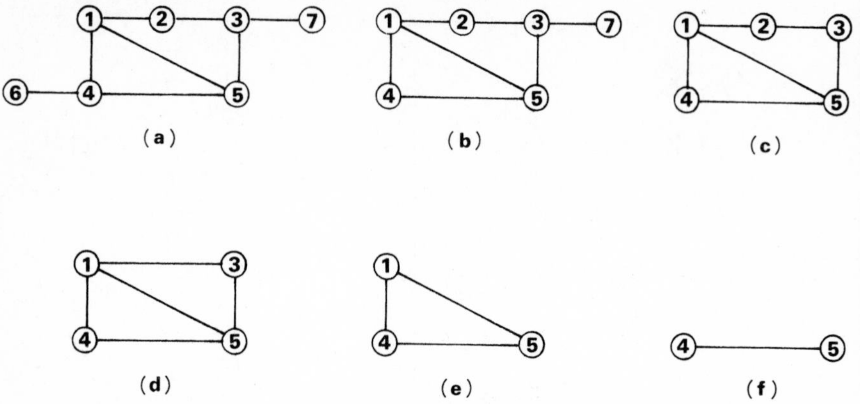

Fig. 5.17 Elimination by minimum number of connected branches in reduced graph. (*a*) original graph; (*b*) after first reduction; (*c*) after second reduction; (*d*) after third reduction; (*e*) after fourth reduction; (*f*) after fifth reduction

can be illustrated by considering the sparse matrix whose graph was previously shown in Fig. 5.10 and is now shown in Fig. 5.17(*a*).

In the original graph there is only one branch connected to nodes 6 and 7. Selecting arbitrarily node 6 for the first elimination creates the reduced graph shown in Fig. 5.17(*b*). After the first reduction, node 7 has the least number of connected branches and therefore is selected for the next elimination. This creates the reduced graph shown in Fig. 5.17(*c*). After the second elimination, nodes 2, 3 and 4 have the least number of branches connected to them. Selecting arbitrarily node 2 for the next elimination creates the reduced graph shown in Fig. 5.17(*d*). Nodes 3 and 4 now have the least number of branches and selecting arbitrarily node 3 creates the reduced graph shown in Fig. 5.17(*e*). Selecting node 1 from this reduced graph creates the graph shown in Fig. 5.17(*f*). Finally, the process is completed by eliminating nodes 4 and 5 in any order. Therefore, although it is evident that several sequences are possible, one order of elimination is:

(6,7,2,3,1,4,5)

The method of deciding the most suitable node to be eliminated at each reduction step can, as described in section 5.4, be achieved by inspecting the corresponding reduced matrix. At each stage of elimination, the diagonal element of the reduced matrix in the row and column which has the least number of off-diagonal elements is selected as the next pivot; if more than one diagonal element satisfies this criterion, any of them can be selected. It is evident that this inspection technique is a fairly simple one to program. The number of columns or rows and the number of elements in each column or row which must be inspected decreases as the elimination progresses. Therefore this method is computationally efficient and achieves an ordering which is close to the optimum.

5.5.2 Introduction of least number of new branches

In this scheme, the node of the reduced graph which, when eliminated, introduces the least number of new branches, is selected as the most suitable node for elimination. If more than one node has this property, any of them may be selected. This technique, based on valency, minimizes the amount of fill-in of the reduced coefficient matrix at each reduction step. This process of selection is not necessarily the same as selecting a node with the minimum number of branches connected to it.

To illustrate this ordering scheme, consider the previous example, which is redrawn in Fig. 5.18(*a*). Using the principle of least number of new branches, the same nodes (6 and 7) will be selected for elimination at the first two steps, since these do not introduce any new branches in the subsequent reduced graphs. At the third step, however, the schemes differ.

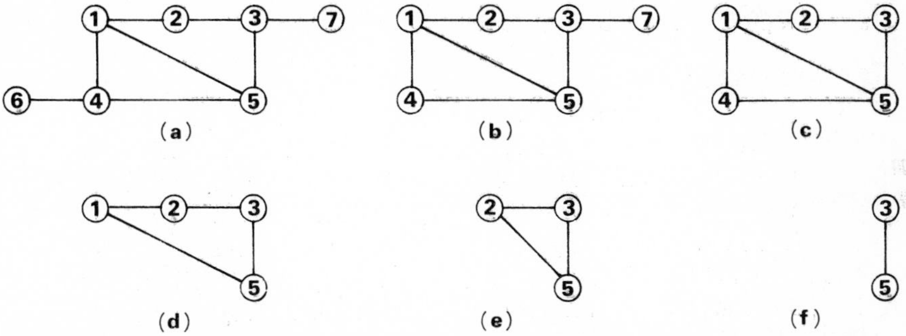

Fig. 5.18 Elimination by minimum number of new branches in reduced graph. (*a*) original graph; (*b*) after first reduction; (*c*) after second reduction; (*d*) after third reduction; (*e*) after fourth reduction; (*f*) after fifth reduction

Although nodes 2, 3 and 4 each have two branches connected to them, eliminating node 4 does not introduce any new branches because nodes 1 and 5, which are directly connected to node 4, are already interconnected. Eliminating node 2, however, would introduce a new branch between nodes 1 and 3, and eliminating node 3 would introduce a new branch between nodes 2 and 5. Similarly, if either node 1 or node 5, which have three branches connected to them, is eliminated, two new branches are introduced. Therefore, for the third reduction step, node 4 is selected for elimination, which creates the reduced graph shown in Fig. 5.18(*d*). Eliminating any of the four remaining nodes introduces one new branch. Therefore, selecting arbitrarily node 1 for elimination creates the graph shown in Fig. 5.18(*e*). Eliminating any of the three remaining nodes does not introduce any new branches, and node 2 can be selected, creating the reduced graph shown in Fig. 5.18(*f*). Finally, the process can be completed by eliminating nodes 3 and 5 in any order. It is again evident that several sequences are possible, one order of elimination being:

(6,7,4,1,2,3,5)

This method is more difficult to program than the previous dynamic ordering technique and in terms of time is computationally less efficient. However, it does minimize the fill-in at each reduction step and therefore decreases the number of non-zero elements created during elimination and reduces the required storage.

This method can be extended so that, at each reduction step, it selects for elimination a node which creates the least number of new branches for two steps ahead. It may be further extended to include several steps ahead or, in the limit, to survey the total number of reduction steps necessary to eliminate all the nodes of the network. This complete extension achieves maximum conservation of sparsity and therefore the optimal ordering is obtained. The extra computation necessary to achieve this ideal may, however, be prohibitive for most practical applications and consequently the method is not generally used.

5.6 Comparison of ordering schemes

Two fundamental factors must be considered when deciding which method would be the most suitable for solving a given set of large, sparse simultaneous linear equations by digital computer. These are the storage requirement and the computation time. Any consideration in respect of these factors must be dependent on the computing facilities available. Even with very fast modern computers having large core stores, it is still desirable to use a method which requires minimum storage and computation time; the increase in computer size enables problems of greater size and complexity to be solved with little increase in computational effort. As the available computing capacity decreases, it becomes even more imperative to use the modern efficient computing techniques.

Several possible ordering techniques have been discussed in the previous sections of this chapter, all of which can be used in conjunction with any of the factorization methods discussed in Chapter 4. There is, however, no one technique which is universally applicable, and it is often necessary to compromise between several desirable goals.

Network-type problems, as discussed in chapter 1, occur in a very wide and diverse range of activities, some of which are in the engineering field, such as power-system design, and others in non-engineering activities, such as accountants' cash-flow problems. These problems, although characterized by the common feature of a high degree of sparsity, can have very different sizes and sparsity structures. In consequence, a different type of ordering technique may be desirable. Also, as briefly discussed in the previous sections of this chapter, one problem may be a one-off calculation, whilst another may involve many similar repetitive calculations. This aspect can be an additional pertinent consideration in deciding which method is the most suitable.

The following discussion attempts to rationalize some of these considerations and to indicate how a decision may be formulated. It is, of course, impossible to cover all of the possible combinations of problems, problem size and problem sparsity structure, and therefore the discussion can only be made in general terms. The points discussed, however, when put into the context of the problem to be solved, should enable a realistic decision to be made.

For small problems involving only a few variables, there are only marginal differences between the various methods. In fact, if the problem is very small, having up to, say, about 10–15 variables, little time is lost by using direct matrix inversion. This method may involve significantly increased storage, but as the problem is small, this increase is of little consequence. As the size of the problem increases, the differences between the various methods become more apparent. For problems involving, say, 50 or more variables, it is worthwhile to consider not only the best ordering method but also the structure and efficiency of the complete program.

At this juncture it is worth noting that if small problems are the only ones to be solved, then, unless an existing program is available, little is gained by attempting to program a complex ordering and factorization method, since the time saved in execution can be considerably outweighed by the time spent in programming and debugging. If small and large problems having a similar structure are to be solved, however, then it is very desirable to use a program based on efficient ordering and factorization.

In comparing the various ordering methods, it is essential to consider not only the efficiency of the ordering process itself but also how this affects the factorization and solution efficiency. For instance, a very fast ordering process may produce very inefficient and computationally slow factorization and solution. It is in arriving at a reasonable compromise between these two aspects that the structure of the problem can play a very important part. Again, these points are mentioned as appropriate in the ensuing discussion.

In general, the pre-ordering schemes are simple and computationally very fast but are usually rather poor in preserving sparsity, and a relatively large amount of fill-in normally occurs. The required storage therefore becomes significant and the computational time of factorization and numerical solution increases quite considerably. For large systems, the increase in storage and solution time normally outweighs the advantage of fast ordering, and, in general, pre-ordering is not a satisfactory method for such problems. It can, however, be justified in two specific cases. The first is if the problem is relatively small and particularly if of the radial type. The second is if the problem, although large, is a one-off problem. In this case, the loss of computational efficiency can outweigh the effort spent in programming a more efficient method.

The banding schemes are very restricted in their application and are only useful for special types of problems. They do, however, have several advantages, which probably accounts for their popularity. The elimination process is systematic, which is useful with a backing store. Also, the required storage is determined in advance by the bandwidth. Finally, being a pre-ordering process, the banding scheme is very simple to program. On the other hand, the scheme has several important disadvantages. When the elimination and numerical solution times are incorporated in their overall appraisal, these methods are found to be inefficient for general network problems. Also, although the band may be sparse in many problems, it is likely that it will be completely filled after elimination. A great deal of work has been done to improve their efficiency, particularly in connection with minimizing the required bandwidth. Even so, modern banding techniques appear to be unattractive for general network analysis and in cases where the band is wide relative to the order of the matrix.

For the general type of large, sparse network-type problems, a form of dynamic ordering seems to be the only really satisfactory method, particularly as the problems increase in size and if several problems of the same type are to be solved. The dynamic ordering process is not easy to program and takes significantly longer to execute than any of the pre-ordering schemes. However, when the factorization and numerical solution efficiency are incorporated into the overall appraisal, these methods immediately become very attractive.

The first method discussed in section 5.5 and based on minimum number of connected branches is much simpler and faster than the alternatives. For the general type of network problems, it creates near-optimum conservation of sparsity. This method is the easiest of its type to program and is the most practical one for general network problems. The dynamic ordering method, based on valency, is more difficult to program and is more time-consuming to execute. It can, however, achieve an improved preservation of sparsity, although this gain does not seem to be significant in most practical cases. The extension of the valency method to achieve optimal ordering does not seem to offer any significant increase in preservation of sparsity and takes a much longer time to execute. Consequently optimal ordering techniques do not seem to be attractive or practical for general large, sparse networks.

The authors have tried many of the available techniques for ordering and factorization. After some considerable time and experience they are firmly of the belief that for general large, sparse network problems, a combination of factorization and simple dynamic ordering is probably the best available compromise. This combination has been applied to a wide range of diverse problems, particularly in the engineering field, and very efficient and accurate solutions have always been obtained. They do accept, however, that in certain circumstances and with special problems a different combination may be preferred. In such cases, they hope that the foregoing discussion will be of considerable assistance in selecting a suitable combination of methods.

5.7 Network decomposition

In the preceding sections of this chapter, the sparsity-directed ordering methods have been discussed assuming that the whole system is analysed simultaneously, that is, the associated coefficient matrix or graph of the complete network is reduced sequentially. There are, however, additional techniques which can be used either on their own or as part of the preceding techniques.

These additional techniques are important for problems in which sparsity techniques alone are not sufficient for an efficient solution. This can occur in practical problems which contain groups of highly interconnected nodes, the individual groups being only sparsely connected. An example of this exists in power transmission and distribution systems. Each densely populated or industrial area will have a number of nodes interconnected by many lines, whereas the different areas will only be connected by a few lines.

For problems of this type, two degrees of sparsity exist: one within each sub-system and the other between the different sub-systems. Sometimes it is advantageous to solve such problems by partitioning the coefficient matrix into blocks, each block corresponding to one of the more closely interconnected sub-systems. Alternatively, the network itself could be decomposed into clusters of sub-systems. Certain advantages can sometimes be gained by such decomposition and occasionally it even becomes a necessity. Typical examples of such occasions are:

(i) The solution of a large, sparse problem using a time-sharing computer. This may be solved more efficiently by first decomposing it into sub-systems. This allows a more efficient use of the backing store by simultaneously doing calculations and transfers.

(ii) Controlling a large integrated system, using a number of small but fast local computers linked to a master computer. Decomposition of the system is essential for this type of control operation.

(iii) The making of relatively small but frequent changes to part of a given sub-system, for example, switching operations of power system transmission or distribution feeders. Solution of these problems can be obtained more efficiently by re-solving for the sub-system changes and superimposing the effect on the previous solution.

(iv) When, in certain problems, decomposition can create two or more sub-systems which are identical. This achieves savings in both the storage and computation time.

There are two fundamentally different approaches in solving problems of this type: the matrix method and the diakoptical approach. In the matrix method, the coefficient matrix is partitioned into blocks of sub-matrices or alternatively the blocks are constructed from a knowledge of the topology or graph of the problem. In the diakoptical approach, the system is subdivided by removing or 'tearing' some of the interconnections and each part is solved independently using any orthodox method, for example, sparsity. The component solutions are then superimposed using a simple systematic process to obtain the solution of the original system. In general, less storage and computation time is required for the diakoptical approach, although the matrix partitioning method is simpler to implement.

The overall computational efficiency and the amount of storage required with both methods depends on the way the system is decomposed. Generally the efficiency is increased if the tearing or partitioning of the system produces sub-systems which have a minimum amount of coupling between them.

Engineers can identify sub-systems relatively easily by inspection of a good layout of the network. The effort required to identify and separate the sub-systems using a digital computer is comparatively much greater. There are some existing schemes which will identify the degree of interconnection between the various nodes and enable the sub-systems to be separated. These schemes require the enumeration of all the potential grouping pairs and are only practical for relatively small networks. Approximate methods appear to be more promising for the larger systems, but even these may not be very practical for systems with more than about 100 nodes. There is, however, a considerable general interest in these techniques and it is envisaged that some practical schemes will be proposed in the near future. In the meantime it is preferable that network decomposition should be made on geographical considerations and the judgement and experience of the engineers and system analysts.

Bibliography

Akyuz, F. A. & Utku, S.: An automatic relabelling scheme for bandwidth minimization of stiffness matrices. *Am. Inst. Aeronaut. Astronaut. J.* **6**, 728–30, 1968.

Allway, G. G. & Martin, D. W.: An algorithm for reducing the bandwidth of a matrix of symmetrical configuration. *Comput. J.* **8**, 264–72, 1965.

Arany, I., Smyth, W. F. & Szoda, L.: An improved method for reducing the bandwidth of sparse, symmetric matrices. *Proc. Int. Fedn Inf. Processg Conf.*, Ljubljana, 1971.

Berry, R. D.: An optimal ordering of electronic circuit equations for a sparse matrix solution. *Inst. elect. electron. Engrs Trans.* (CT-18), **1971**, 40–50.

Brameller, A. & Allan, R. N.: The role of sparsity in the analysis of large systems. *Comput. Aided Des.* **6**, 159–68, 1974.

Brameller, A., John, M. N. & Scott, M. R.: *Practical Diakoptics for Electrical Networks.* Chapman & Hall, 1969.

Brameller, A. & Lo, K. L.: The application of diakoptics and the escalator method to the solution of very large eigenvalue problems. *Int. Jl numl Meth. Engng* **2**, 535–49, 1970.

Cantin, G.: An equation solver of very large capacity. *Int. Jl numl Meth. Engng* **3**, 379–88, 1971.

Carpentier, J.: Ordered eliminations. *Proc. Power Syst. Computn Conf.*, London, 1963.

Chang, A.: Application of sparse matrix methods in electric power system analysis. In *Sparse Matrix Proceedings*, pp. 113–21. IBM, 1969.

Edelmann, H.: Ordered triangular factorisation of matrices or multi-purpose approach for treating problems in large electrical networks. *Proc. Power Syst. Computn Conf.*, Stockholm, 1966.

Hansen, J. R.: A new procedure for topologically controlled eliminations. *Proc. Power Syst. Computn Conf.*, Grenoble, 1972.

Happ, H. H.: *Diakoptics and Networks.* Academic Press, 1971.

Harary, F.: Sparse matrices and graph theory. In *Large Sparse Sets of Linear Equations*, pp. 139–50. Academic Press, 1971.

Jennings, A. & Tuff, A. D.: A direct method for the solution of large sparse symmetric simultaneous equations. In *Large Sparse Sets of Linear Equations*, pp. 97–104. Academic Press, 1971.

King, I. P.: An automatic re-ordering scheme for simultaneous equations derived from network analysis. *Int. Jl numl Meth. Engng* **2**, 523–33, 1970.

Kron, G.: *Diakoptics.* Macdonald, 1962.

McCormick, C. W.: Application of partially banded matrix methods to structural analysis. In *Sparse Matrix Proceedings,* pp. 155–8. IBM, 1969.
Nathan, A. & Even, R. K.: The inversion of sparse matrices by a strategy derived from their graphs. *Comput. J.* **10**, 190–4, 1967.
Norin, R. S. & Pottle, C.: Effective ordering of sparse matrices arising from non-linear electrical networks. *Inst. elect. electron. Engrs Trans.* (CT-18), **1971**, 139–45.
Ogbuobiri, E. C., Tinney, W. F. & Walker, J. W.: Sparsity directed decomposition for Gaussian elimination on matrices. *Inst. elect. electron. Engrs Trans.* (PAS-89), **1970**, 141–50.
Parter, S.: The use of linear graphs in Gauss elimination. *Symp. appl. Math. Rev.* **3**, 119–30, 1961.
Rose, D. J.: Triangulated graphs and the elimination process. *J. math. Anal. Applic.* **32**, 597–609, 1970.
Rosen, R.: Matrix bandwidth minimisation. *Proc. 23rd Natn. Conf. Ass. Computg Mach.* (Publ. P-68), pp. 585–95. Brandon Systems Press, 1968.
Roth, J. P.: An application of algebraic topology: Kron's method of tearing. *Q. appl. Math.* **17**, 1–24, 1959.
Sato, N. & Tinney, W. F.: Techniques for exploiting the sparsity of the network admittance matrix. *Inst. elect. electron. Engrs Trans.* (PAS-82), 944–50, 1963.
Spillers, W. R.: Analysis of large structures: Kron's method and more recent work. *J. struct. Div., Proc. Am. Soc. civ. Engrs* **94**, 2521–34, 1968.
Spillers, W. R. & Hickerson, N.: Optimal elimination for sparse symmetric systems as a graph problem. *Q. appl. Math.* **26**, 425–32, 1968.
Stott, B. & Hobson, E.: Solution of large power system networks by ordered elimination; a comparison of ordering schemes. *Proc. Instn elect. Engrs* **118**, 125–34, 1971.
Tinney, W. F.: Comments on sparsity techniques for power system problems. In *Sparse Matrix Proceedings,* pp. 25–34. IBM, 1969.
Tinney, W. F. & Walker, J. W.: Direct solutions of sparse network equations by optimally ordered triangular factorisation. *Proc. Inst. elect. electron. Engrs* **55**, 1801–9, 1967.
Zollenkopf, K.: Bi-factorisation, basic computational algorithm and programming techniques. In *Large Sparse Sets of Linear Equations,* pp. 75–96. Academic Press, 1971.

6 Sparsity-directed Programming

6.1 Introduction

It was shown in the preceding chapters that the network type of problem is often characterized by a significant degree of sparsity. The simultaneous equations describing the behaviour of these networks can be solved by direct matrix inversion. Although these methods are fairly easy to program, they cannot, however, exploit sparsity, and unfortunately produce a completely full inverse matrix. For large problems, the storage is therefore extremely large and the methods are very inefficient. Consequently these methods are only suitable for very small problems or problems which have a completely full coefficient matrix.

The alternative direct methods to matrix inversion are the factorization techniques based on Gauss elimination. With these methods sparsity can be exploited and, with a suitable ordering technique, a direct solution can be obtained with a minimum amount of storage and computation time.

These factorization methods can be used to exploit sparsity in two ways. Firstly, by using them in conjunction with a suitable ordering technique, the number of new non-zero elements produced during factorization can be minimized. Secondly, these methods only need the non-zero elements to be stored and processed. Because only the non-zero elements of the original and reduced coefficient matrices are stored in a digital computer, they are clearly not arranged in an array that is visually identical to the corresponding matrix. Instead they are stored in what is generally called a *packed* or *compact* form. It is therefore necessary to locate each of these non-zero elements during the factorization process and identify it as a particular element of the corresponding coefficient matrix. This requires indexing information, in addition to the numerical value of the non-zero element, to be incorporated and stored, so that the relative position of each non-zero element can be established.

In addition to the original non-zero elements, it was shown in Chapter 5 that new non-zero elements are continually being generated during the factorization process. Also, elements which were previously non-zero, may become zero. It is evident that the compacting and indexing schemes must be capable of implementing efficiently these continuous changes during factorization by incorporating the new non-zero elements in store and recording the available storage space.

Factorization methods are, in themselves, relatively easy to program, since the technique of each method is based solely on the Gauss elimination process. The simplicity of these methods, however, is upset when the coefficient matrix as a whole is not stored but only the non-zero elements in a compact form. This compacting requires additional programming and execution time simply to identify each element. It is this aspect of the overall problem that can make one program very efficient and another program very inefficient, although both may be based on the same factorization and ordering methods.

It is not easy to define quantitatively matrices which are amenable to sparsity techniques and programming, as this is dependent not only on the order of the matrix and the number of non-zero elements but also on the relative position of these non-zero elements in the coefficient matrix. For very small problems, the extra effort required to exploit sparsity may be considered not worthwhile. With large problems, having an order of about 50 and above, however, sparsity directed elimination and programming is generally worth the effort involved, particularly if the non-zero elements are distributed relatively uniformly through the matrix. For sparse problems of about 1000 variables or more, these techniques are essential. If both small and relatively large problems occur, then the sparsity-directed program can and should be used also for the small problem.

It is clear from the foregoing that the problems of sparsity-directed programming are associated not only with the basic aspects of factorization and numerical solution but also with the problem of storing and identifying the non-zero elements. This chapter is therefore intended to describe some of the various schemes which the authors consider to be both practical and efficient not only for programming the numerics of the problem but also for programming the addressing of the non-zero elements.

6.2 Storage of a list of numbers

Before describing techniques for storing all the non-zero elements of a sparse matrix in compact form, it is useful to discuss the general aspects involved in simply storing and identifying a series of numbers in a digital computer and the problems associated with modifying this storage when new numbers are to be added to the store. In any computer program, this problem requires the definition of a number of arrays in which the elements and their identification are stored and from which they can always and easily be recalled.

To illustrate the techniques involved, consider the following list of numbers, which it is desired to store in a digital computer:

30·5, 50·9, 26·3, 45·7

This list of numbers could be stored in an array defined as VALUE in the same sequence as they are given above. In this case, the numbers in the array will be:

Location	1	2	3	4	5	6
VALUE	30·5	50·9	26·3	45·7	–	–

There are many other sequences in which the same list of numbers can be stored. Suppose, for instance, that these numbers are to be stored in ascending numerical order. One way of achieving this requirement is to change the numerical value of the elements in each location. Doing this, the new numbers in the array VALUE would be:

Location	1	2	3	4	5	6
VALUE	26·3	30·5	45·7	50·9	–	–

So far this process is very straightforward, even if a re-ordering routine has to be included in the program to arrange the numbers in a certain preferred sequence. Suppose now that it is necessary to add a new number to the list. If the new number can be added to the end of the list then the process is very simple. The size of the list is increased by one and the new number is inserted in the extra available position. If, however, the number is to be inserted in the middle of the existing list, then the process is more complicated.

Let the new number to be added be 28·2 and the new list to be kept in an ascending numerical order. One way of achieving this is to move the last value one position down the list, move the last but one value to the position previously occupied by the last value and continue this process until the new value can be inserted into the appropriate position. Using this method, the new list with the value 28·2 added becomes:

Location	1	2	3	4	5	6
VALUE	26·3	28·2	30·5	45·7	50·9	–

For a long list, this process of inserting new numbers in a particular position within an existing list can be computationally very time-consuming. A more efficient method is to use a technique known as a *linked list.* Linked lists enable the numerical values of the numbers to be stored in any order, the desired sequence of the numbers being determined by the linking technique. This linking technique consists of allocating a storage location for the numerical value of each item of a list and associating with this storage location the address for the numerical value of the next item. This associated address is shown for one location in Fig. 6.1.

numerical value of an item
address of next item ≠ 0 if more items to follow = 0 if present item is last item

Fig. 6.1 One item of a linked list

This linking technique requires the introduction of a new array which may be called NEXT and in which the address of the next required number in the list is stored. Using such an array, the list of the numbers in the original sequence is:

Location	*1	2	3	4	5	6
VALUE	30·5	50·9	26·3	45·7	–	–
NEXT	2	3	4	0		

In addition to the above list, it is necessary to record the address of the first number in the list. This can be stored in practice as a single integer variable, but for simplicity in the above example this element is marked by an asterisk.

To change the order of the list into ascending numerical values, the sequence of numbers in the array VALUE can be left undisturbed and the order modified by changing the addresses in the array NEXT. This gives:

Location	1	2	*3	4	5	6
VALUE	30·5	50·9	26·3	45·7	–	–
NEXT	4	0	1	2		

If now the number 28·2 is to be added in the list and the sequence of numbers is to remain in ascending numerical order, it is sufficient to add this new number to the end of the existing list and to change the addresses in the array NEXT. This gives:

Location	1	2	*3	4	5	6
VALUE	30·5	50·9	26·3	45·7	28·2	–
NEXT	4	0	5	2	1	

Using the array NEXT to re-arrange the sequence of numbers in order to accommodate a new number within the list is computationally considerably more efficient than the shuffling process that had to be performed with the numbers in the array VALUE in the previous technique. Previously all the numbers greater than the new value had to be shifted sequentially, which, with long lists, could involve a considerable number of operations. In the present technique only one value in the original NEXT has to be changed and one new value added. The merits of this scheme are therefore clearly evident. It does, however, require additional storage and this must be balanced against the reduced computation time.

The examples of linked lists described above may appear trivial. They are, however, the basis of one of the most efficient and powerful techniques for storing large, sparse matrices in a compact form.

6.3 Storage of sparse matrices

In the preceding chapters of this book, it has been stressed several times that only the non-zero elements of a sparse matrix need to be stored in a digital computer. Since the storage array will not be identical to the original matrix, both the values of the non-zero elements and a means of identifying these elements must be incorporated in the storage technique.

In the techniques described in section 6.2, the list of numbers being stored formed a vector and only one simple identification label was required. In the storage of a matrix, however, the identification becomes more complex because it must recognize not only the rows but also the columns. There are a number of practical schemes which achieve this objective. Three of these schemes are described in the following subsections. These schemes are illustrated by considering the storage in a compact form of the following sparse symmetrical coefficient matrix **A**:

$$\mathbf{A} = \begin{bmatrix} 3{\cdot}0 & -1{\cdot}0 & -1{\cdot}0 & \cdot \\ -1{\cdot}0 & 2{\cdot}0 & \cdot & \cdot \\ -1{\cdot}0 & \cdot & 2{\cdot}0 & -1{\cdot}0 \\ \cdot & \cdot & -1{\cdot}0 & 1{\cdot}0 \end{bmatrix}$$

6.3.1 Scheme I

As discussed above, the row and column position of a given element must be designated in order to identify the relative position of the element in the matrix. This can be achieved easily and simply by storing all the relevant information of each element in three arrays. These can be defined as VALUE (numerical value of the element), IROW (index of row) and ICOL (index of column). This scheme is therefore very simple, since the location of each non-zero element is defined by the 'co-ordinates' of its position in the matrix. Using this technique, the storage of matrix **A** can be as follows:

Location	1	2	3	4	5	6	7	8	9	10
VALUE	3·0	−1·0	−1·0	−1·0	2·0	−1·0	2·0	−1·0	−1·0	1·0
IROW	1	1	1	2	2	3	3	3	4	4
ICOL	1	2	3	1	2	1	3	4	3	4

Using this method of storage, all the non-zero elements are stored and each is identified by indices in IROW and ICOL. This basic method can, in certain circumstances, be modified to reduce the amount of storage required. For instance, if all the diagonal elements of the matrix are non-zero – the normal case for most network-type problems – a reduction in storage can be achieved by storing the diagonal elements separately. In such cases, no indexing information is required, since the elements can be stored in their natural order and their relative positions are known implicitly. These elements can therefore be stored in a single array, for example, DIAG, as shown below. Also, a further storage saving can be achieved with symmetrical matrices by storing only the upper (or lower) triangular matrix; the unstored elements being known implicitly. Using these two modifications of the basic scheme, the matrix **A** can now be stored in the following form:

Location	1	2	3
VALUE	−1·0	−1·0	−1·0
IROW	1	1	3
ICOL	2	3	4

(*a*) upper off-diagonal elements

Location	1	2	3	4
DIAG	3·0	2·0	2·0	1·0

(*b*) diagonal elements

Comparing the storage requirements using the basic method with that using the modified version, it is seen that the amount of storage required has been reduced from 30 positions to 13.

Even with asymmetrical matrices a simplification of the basic method can be made, provided such matrices have incidence symmetry. In these cases, it is sufficient to add an extra array to that shown in (*a*) above. This additional array, ASVAL say, gives the values of the elements below (or above) the diagonal and uses the index arrays IROW and ICOL in an opposite sense, that is, values in IROW are read as column indices and values in ICOL as row indices. These arrays are therefore located as shown below:

Location	–	–	–	–
VALUE	–	–	–	–
IROW	–	–	–	–
ICOL	–	–	–	–
ASVAL	–	–	–	–

which for an asymmetrical matrix with incidence symmetry has a smaller storage requirement than the original basic method.

6.3.2 Scheme II

In Scheme I, two index arrays were required to identify each of the non-zero off-diagonal elements. This second scheme enables these arrays to be reduced to one without losing any indexing information.

With each non-zero element VALUE (l) of a given matrix it is possible to associate a unique integer IPOS (l), from which the row and column position of the element can be determined. If an element a_{ij} in row i and column j of a matrix **A** having an order n is stored in location l, then the index value can be obtained from:

$$\text{IPOS}\ (l) = i + (j - 1)n$$

and this single index value identifies both the row and column position of the element. The required storage therefore consists of two arrays: VALUE (numerical value of the element) and IPOS (index of position). Using this scheme, the matrix **A** can be stored as:

Location	1	2	3	4	5	6	7	8	9	10
VALUE	3·0	−1·0	−1·0	−1·0	2·0	−1·0	2·0	−1·0	−1·0	1·0
IPOS	1	5	9	2	6	3	11	15	12	16

This can again be considered the basis of the scheme and, as with scheme I, for symmetrical matrices with non-zero diagonal elements can be reduced to the following form:

Location	1	2	3
VALUE	−1·0	−1·0	−1·0
IPOS	5	9	15

Location	1	2	3	4
DIAG	3·0	2·0	2·0	1·0

To reconstruct the row and column position of any element stored in location l, the following computational technique must be used.

The column position j is obtained from the index value by:

$$j \text{ is the smallest integer} \geqslant \frac{\text{IPOS}\ (l)}{n}$$

and the row position i is obtained from:

$$i = \text{IPOS}\ (l) - (j - 1)n$$

For example, IPOS (3) = 15,

$$\frac{\text{IPOS}\ (3)}{n} = \frac{15}{4}$$

and the smallest integer greater or equal to $\frac{15}{4}$ is 4.

Therefore $j = 4$

and
$$\begin{aligned} i &= \text{IPOS}\ (3) - (j - 1)n \\ &= 15 - (3 \times 4) \\ &= 3 \end{aligned}$$

and the element in location 3 is in the third row and fourth column of the matrix, i.e. element a_{34}.

The advantage of this scheme is that the storage is less than that required with scheme I. This advantage is gained, however, at the expense of the additional computation time required to calculate the index values and to reconstruct the row and column positions from these index values. Therefore, neither scheme can be stated as having complete merit over the other; the decision must rest with the users' required balance between computational time and storage.

6.3.3 Scheme III

Neither of the two previous schemes utilize the efficient linked list technique discussed in section 6.2. The third scheme, however, is based on this technique and, for the reasons discussed previously, is often found to be an extremely useful scheme for storing and manipulating large, sparse matrices. Before describing this scheme it is useful and pertinent to discuss the requirements of a storage scheme in more detail, in order to indicate the reasons why this third scheme can be so advantageous.

In the solution of simultaneous linear equations by any of the Gauss elimination methods, the factorization processes are performed either on a row-by-row basis or on a column-by-column basis. It is therefore very desirable and more efficient to index the stored information in such a way that it can be retrieved during factorization in a similar manner. Also, during the elimination process, non-zero elements are continuously generated or deleted, and therefore the storage scheme should enable these modifications to be made efficiently. Finally, to minimize the generation of new non-zero elements, it is necessary to know the number of non-zero elements in each column (or row). This information may need updating during elimination and therefore the storage scheme should incorporate this information and also be sufficiently flexible to allow pivoting in any desired order.

One possible scheme which satisfies the above requirements is an extension of the linked list storage technique. This storage scheme is discussed below, assuming that the Gauss elimination process is to be performed on a column-by-column basis.

This scheme consists of two tables. The first table comprises three arrays:

VALUE	(numerical value of element)
IROW	(index of row)
NEXT	(location of next non-zero element in the column).

The second table also comprises three arrays:

DIAG	(numerical value of diagonal element)
ICAP	(index of column address pointer)
NOZE	(number of non-zero off-diagonal elements)

These arrays are discussed in more detail below, but first it is useful for this discussion to illustrate how the matrix **A**, used previously, would be stored using this scheme. For this matrix, the two tables and six arrays would be:

Location	1	2	3	4	5	6	* 7	8	9
VALUE	−1·0	−1·0	−1·0	−1·0	−1·0	−1·0	–	–	–
IROW	2	3	1	1	4	3			
NEXT	2	0	0	5	0	0	8	9	10

Location	1	2	3	4
DIAG	3·0	2·0	2·0	1·0
ICAP	1	3	4	6
NOZE	2	1	2	1

In addition to the information stored in the above two tables, it is necessary to know the address of the first free position in the first table. This can be achieved in practice by the use of a single integer variable and has been denoted in the above example by an asterisk.

The meaning and use of some of the arrays shown above will already be evident from the discussion in section 6.2 and the description of the two previous storage schemes; for instance, the arrays VALUE, IROW, NEXT and DIAG have all been used previously and their significance remains unchanged.

Consider first the three elements stored in location l of the first table. VALUE (l) gives the numerical value of a non-zero off-diagonal element, IROW (l) gives the row position of this element and NEXT (l) gives the address of the next non-zero off-diagonal element appearing in the same column of the matrix. A value of zero in the array NEXT indicates that there are no further non-zero elements in that column.

Consider now the three elements stored in location m of the second table. DIAG (m) gives the numerical value of the diagonal element appearing in column m of the matrix, ICAP (m) gives the location in the first table of the first non-zero off-diagonal element in column m and NOZE (m) gives the total number of non-zero off-diagonal elements in column m.

Any column of the matrix can easily be reconstructed from the information stored in these arrays. As an example, consider the reconstruction of column 3.

From the second table, the value of the diagonal element can be found in location 3 of DIAG and is equal to 2·0,

i.e. DIAG (3) = 2·0

Also from the second table, the address in the first table of the first non-zero off-diagonal element in column 3 can be found in location 3 of ICAP and is seen to be 4,

i.e. ICAP (3) = 4

From the first table, the value of the first non-zero off-diagonal element is therefore found in location 4 of VALUE and is equal to $-1\cdot0$,

i.e. VALUE (4) = $-1\cdot0$

The row position of this off-diagonal element is found in location 4 of IROW and is seen to be 1,

i.e. IROW (4) = 1

The address of the next non-zero off-diagonal element in column 3 of the matrix is found in location 4 of NEXT and is seen to be 5,

i.e. NEXT (4) = 5

The value of this off-diagonal element is therefore found in location 5 of VALUE and is equal to $-1\cdot0$,

i.e. VALUE (5) = $-1\cdot0$

Continuing the same technique in a sequential manner, the row position of this element is 4,

i.e. IROW (5) = 4

and the address of the next non-zero off-diagonal element is zero,

i.e. NEXT (5) = 0

and there are no more non-zero elements in column 3 of the matrix.

This reconstruction has therefore shown that column 3 of the matrix is:

$$\begin{bmatrix} -1\cdot0 \\ \cdot \\ 2\cdot0 \\ -1\cdot0 \end{bmatrix}$$

Although the above example may seem trivial and stretched laboriously, this has been done deliberately and normally this scheme would not be applied to such problems. In order to program this scheme efficiently for a large problem it is necessary that a thorough grasp of the principles is acquired, which is easier to achieve by following the above simple example.

The reader will already have realized that the required storage for this scheme appears to be significantly greater than for those discussed previously. This is, however, not strictly correct, since the information stored in this scheme contains all the information that is necessary to reconstruct the matrix from store and to perform the factorization process in any desired sequence of elimination. The two previous schemes do not contain sufficient information if dynamic ordering is required, and therefore additional storage is necessary for such schemes.

The great advantage of the linked list storage technique is that new elements can be added and existing elements deleted with the minimum amount of effort. This was described for a list of numbers in section 6.2. Consider now how this can be implemented in the case of a stored matrix and suppose that two new elements must be added to the existing arrays:

$$a_{14} = -2{\cdot}0 \quad \text{and} \quad a_{41} = -2{\cdot}0$$

This can be done as follows:

The element a_{41} is in row 4 and column 1 of the matrix. Therefore, first locate the last non-zero element in column 1. Using ICAP (1) and the process described previously, this element is found in location 2 of the first table. The new element a_{41} is stored in the first free location of the first table (location 7) and the value of NEXT (2) and NOZE (1) modified accordingly. Therefore, the following modifications are made to the first table:

NEXT (2) = 7
VALUE (7) = −2·0
IROW (7) = 4
NEXT (7) = 0
NOZE (1) = 3

A similar process can be used for storing the element a_{14}, which is added to the first table in location 8.

With these additions, the modified tables become:

Location	1	2	3	4	5	6	7	8	9*
VALUE	−1·0	−1·0	−1·0	−1·0	−1·0	−1·0	−2·0	−2·0	–
IROW	2	3	1	1	4	3	4	1	
NEXT	2	7	0	5	0	8	0	0	10

Location	1	2	3	4
DIAG	3·0	2·0	2·0	1·0
ICAP	1	3	4	6
NOZE	3	1	2	2

Similarly, if an element is to be deleted from the listing, its location becomes available for a new element to be added and the appropriate values in NEXT and NOZE are modified accordingly.

It is evident therefore that the location of the non-zero off-diagonal elements in the arrays need bear no relation to their position in the corresponding matrix, the location and identification being determined only by the addresses in the various index arrays. The diagonal elements, however, are arranged in their natural order.

6.4 Programming principles for ordering and factorization

The schemes for sparsity-directed programming discussed in the previous sections of this chapter can be used in conjunction with any ordering scheme and any factorization method. It is therefore useful to describe the principles involved in the programming of these methods in order to exploit the sparsity structure. With the aid of these principles and the techniques described in this and previous chapters, a competent analyst should be able to create a program to suit his own particular needs and types of problems.

As discussed later, in section 6.7, the authors have found the bi-factorization method used with dynamic ordering to have many applications in the solution of engineering problems. To illustrate the principles of this programming, therefore, this method of solution and ordering will be described below. These techniques, however, are equally applicable to all factorization methods and most of the ordering schemes.

One of the most time-consuming aspects of sparsity-directed solutions is deciding the most suitable order for pivoting. The programming, therefore, is usually a compromise between the overall computation time and storage required.

As described in Chapter 5, the most suitable dynamic ordering for most problems is based on the minimum number of connected branches. This order can be determined by detecting the least number of non-zero elements in any column or row at each reduction step. It would seem most logical for this detection to be performed during the actual factorization process. However, an alternative method would be to determine the complete dynamic ordering sequence before factorization takes place, using a simulated reduction process. This simulation process will clearly be less efficient in computation time if only one set of equations are to be solved. It becomes very efficient, however, if the equations are to be solved for several coefficient matrices having the same sparsity structure but with elements of different numerical values. In this case the same sequence of ordering can be used for all the coefficient matrices, permitting a considerable reduction in overall computation time. It is therefore apparent that this simulation should form a distinct part of the overall program and is best constructed as a separate sub-routine. This aspect will be discussed in more detail in section 6.6, together with the reasons for subdividing the other major parts of the overall program into different sub-routines.

Before discussing the actual factorization process, therefore, we will consider the principles involved in the programming of this simulation. To illustrate these principles, consider the symmetrical coefficient matrix shown below:

$$\mathbf{A} = \begin{bmatrix} 3\cdot0 & -1\cdot0 & -1\cdot0 & \cdot \\ -1\cdot0 & 2\cdot0 & \cdot & \cdot \\ -1\cdot0 & \cdot & 2\cdot0 & -1\cdot0 \\ \cdot & \cdot & -1\cdot0 & 1\cdot0 \end{bmatrix}$$

This matrix can be stored as shown in section 6.3, using scheme III. The requirement of the simulation process is to determine and record the sequence of dynamic ordering based upon minimum number of non-zero elements in any column. Therefore, in addition to the arrays discussed in scheme III, one extra array is required in the second table in which the order of elimination deduced during simulation is recorded. This array may be called NORD (numerical sequence of ordering). The two tables comprising these seven arrays will therefore be as shown below. To separate the storage of the coefficient matrix in compact form from the simulation process and also to be able to re-define new coefficient matrices having the same sparsity structure, it is advantageous to formulate these arrays in a separate sub-routine of the overall program. This aspect will also be discussed in more detail later.

Initially, since no logical steps have been made to determine the order of elimination, the elements in NORD can, if required, be set to zero as shown or set in the natural order, i.e. 1 to 4.

Location	1	2	3	4	5	6	*7	8	9	10
VALUE	−1·0	−1·0	−1·0	−1·0	−1·0	−1·0	–	–	–	–
IROW	2	3	1	1	4	3	–	–	–	–
NEXT	2	0	0	5	0	0	8	9	10	0

Location	1	2	3	4
DIAG	3·0	2·0	2·0	1·0
ICAP	1	3	4	6
NOZE	2	1	2	1
NORD	0	0	0	0

The first step in the simulation process is to find the first pivotal column. This is very simple since it is determined by searching all elements of NOZE (number of non-zero elements in each column) to detect the minimum value. For the particular example being considered, NOZE (2) = 1 and NOZE (4) = 1, indicating that both columns 2 and 4 each have only one non-zero off-diagonal element. Therefore either can be chosen as the first pivotal column. In this example we will choose arbitrarily column 2 and, as this column and the corresponding row will be eliminated first in the actual factorization process, an integer 1 is inserted into NORD (2). The second table therefore becomes:

Location	1	2	3	4
DIAG	3·0	2·0	2·0	1·0
ICAP	1	3	4	6
NOZE	2	1	2	1
NORD	0	1	0	0

The next column to be chosen as pivotal using dynamic ordering depends on the number of non-zero elements which exist in each column of the reduced matrix after column 2 and row 2 have been eliminated. This number of non-zero elements may be different due to new elements being generated during reduction. Therefore, the next step is to simulate the reduction process that would occur during the elimination of column 2 and row 2.

In the simulation process, the numerical values of the non-zero elements are unimportant; the criterion is whether such an element exists. Therefore, the requirement is to find the position only of all the non-zero elements. Prior to the first reduction step, the sparsity structure of the matrix is:

$$\left[\begin{array}{c|c|cc} X & X & X & \cdot \\ \hline X & X & \cdot & \cdot \\ \hline X & \cdot & X & X \\ \cdot & \cdot & X & X \end{array}\right]$$

This matrix also shows the column and row to be eliminated first. By referring to Gauss elimination, the factorization methods or sparsity-directed elimination described previously, the reader will soon verify that, by eliminating column 2 and row 2, no new non-zero elements will be created and only the value of element a_{11} will be changed. Therefore, the structure of the new reduced matrix becomes:

$$\begin{bmatrix} X & \cdot & X & \cdot \\ \cdot & 1 & \cdot & \cdot \\ X & \cdot & X & X \\ \cdot & \cdot & X & X \end{bmatrix}$$

The logical steps involved in this process may be simulated computationally using the following technique:

From the second table, the first element of the pivotal column is in position ICAP (2) = 3, i.e. location 3 of the first table. The row index of this element is IROW (3) = 1, i.e. it is in row 1 of the matrix. Since NEXT (3) = 0, there is only one non-zero off-diagonal element (a_{12}) in column 2, which is, of course, confirmed by the initial decision of NOZE (2) = 1. Because all the remaining off-diagonal elements in the pivotal column (and row) are zero, the first reduction affects the value of element a_{11} only. Since there is already a non-zero element in this position, DIAG (1), this first reduction step does not introduce any new non-zero elements.

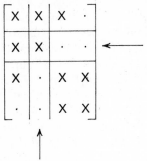

At this point it is necessary to digress slightly, and consider another aspect which is useful to perform during the simulation process. During the actual factorization, the numerical values and indexing of any new non-zero elements must be stored. This can be achieved at the appropriate time by using the linked list designated in the array NEXT. During simulation, however, all new non-zero elements are detected, although their value remains unknown. Therefore, to reduce the factorization time, the allocation of a suitable storage location can be determined simultaneously with the simulation process. This can be computationally more efficient since, during actual factorization, the new non-zero element can be inserted in the predetermined location with no additional effort.

During actual factorization, the non-zero elements of the pivotal column and row are eliminated from the matrix, as shown in the reduced matrix above; the relevant information of this column and row is incorporated in one of the transformation or factor matrices. In terms of the computational storage scheme, the elements of the factor matrices can be stored in the location previously occupied by the corresponding element of the coefficient matrix. However, for symmetrical matrices, as being considered in this example, only the left-hand factor matrices produced during bi-factorization need be stored. Also, since each corresponding column and row of the coefficient matrix are identical, the left-hand factor matrix can be determined completely from either the column or the row; consequently one, in this example the row, can be deleted from store and the vacated location used to accommodate a new non-zero element.

From these slight digressions it is now possible to explain how the first and second tables are modified to simulate the reduction process.

Firstly, consider the first table and remember that column 2 was being used as the pivotal column. By searching through the array IROW, all the non-zero elements in row 2 can be deleted. In this case only one element is involved, in location 1. In practice, the values would not be deleted but simply overwritten as and when required. However, to illustrate the principles involved, the values have been removed from the new first table shown below. Location 1 is now vacant and can be used to accommodate a new non-zero element. The free position locater (*) is moved to location 1. Also, the value of NEXT (1) is set equal to 7, the previous first-available free location. All the remaining values in the first table remain unchanged.

Now consider the second table. Since the elements of row 2, i.e. location 1 of the first table, have been deleted, the index in ICAP (1) must be modified. This must now refer to the next non-zero element in column 1, i.e. the element stored in location 2 of the first table. This can be determined by using the index in NEXT (1), and NEXT (1) = 2. The value of ICAP (1) is therefore set to 2. If there were no further non-zero elements in column 1, NEXT (1) would have been 0 and therefore ICAP (1) would have also been set to 0. Similarly, since simulation has detected that one non-zero element is deleted from column 1, the value of NOZE (1) must also be modified, this now being NOZE (1) = 1.

With these modifications, achieved computationally, the two new tables become:

Location	*1	2	3	4	5	6	7	8	9	10
VALUE	–	−1·0	−1·0	−1·0	−1·0	−1·0	–	–	–	–
IROW	–	3	1	1	4	3	–	–	–	–
NEXT	7	0	0	5	0	0	8	9	10	–

Location	1	2	3	4
DIAG	3·0	2·0	2·0	1·0
ICAP	2	3	4	6
NOZE	1	1	2	1
NORD	0	1	0	0

The above process can now be repeated to determine the next column and row to be eliminated and to simulate the reduction step. Searching NOZE in the new second table shows that either column 1 or column 4 can be used as the pivotal column. Choosing column 1 as pivotal causes the matrix to be reduced as shown below:

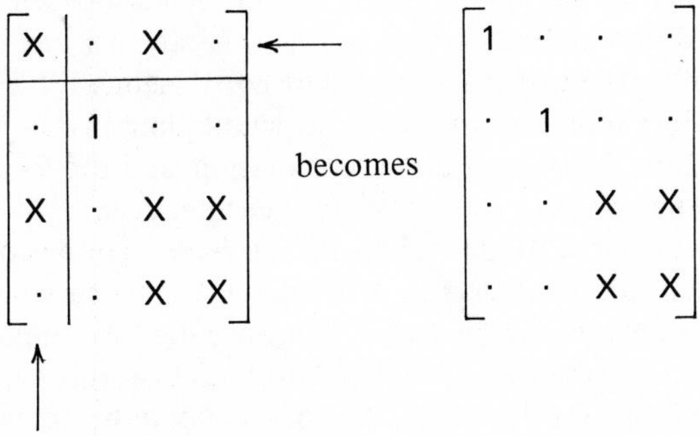

Using the computational principles described for the first reduction step, this second reduction can be simulated, which produces the two new tables shown below:

Location	1	2	3	*4	5	6	7	8	9	10
VALUE	–	−1·0	−1·0	–	−1·0	−1·0	–	–	–	–
IROW	–	3	1	–	4	3	–	–	–	–
NEXT	7	0	0	1	0	0	8	9	10	–

Location	1	2	3	4
DIAG	3·0	2·0	2·0	1·0
ICAP	2	3	5	6
NOZE	1	1	1	1
NORD	2	1	0	0

Repeating the above procedure once again, this time using column 3 as the pivotal column, gives for the third simulated reduction:

Location	1	2	3	4	5	* 6	7	8	9	10
VALUE	–	−1·0	−1·0	–	−1·0	–	–	–	–	–
IROW	–	3	1	–	4	–	–	–	–	–
NEXT	7	0	0	1	0	4	8	9	10	–

Location	1	2	3	4
DIAG	3·0	2·0	2·0	1·0
ICAP	2	3	5	0
NOZE	1	1	1	0
NORD	2	1	3	4

In the latest second table shown above, the value of NORD (4) can be set to 4 directly, since column 4 will be the last column to be selected as a pivotal column. Also, the value of ICAP (4) is found to be zero, which indicates that, after the third reduction, there are no non-zero off-diagonal elements in column 4. This is confirmed by the value of NOZE (4) which is also zero. Therefore, the fourth and final reduction step cannot generate any new non-zero elements and can only affect the value of the diagonal element. Consequently the two latest tables represent the complete simulation of the factorization process and the dynamic order of elimination, and the final number and position of the non-zero elements have been deduced.

In this simulation process, no arithmetical operations have been performed, only a series of logical steps using integers to determine the most suitable order of elimination. This process can be achieved, using a digital computer, in a much shorter time than is required for arithmetical calculations.

In the previous method it was assumed that the order should be determined by detecting only one column having the minimum number of non-zero elements at each reduction step and ignoring all other columns.
Reduced computational effort is achieved if, at a given reduction step, all the columns having the same minimum number of non-zero elements are selected and placed in natural ascending order. For instance, the first set of tables shown on page 112 indicates that columns 2 and 4 each have one non-zero element. The previous method selected only column 2 and proceeded with the reduction process. This alternative method would select column 2 as the first pivotal column and column 4 as the second pivotal column. This technique is known as group elimination.

Let us now consider the principles involved in extending this simulation process to the actual factorization of the matrix. This becomes relatively simple using the order of elimination determined during simulation and processing only the non-zero elements. In order to re-solve the basic equations with different coefficient matrices having the same sparsity structure using the sequence of elimination deduced during simulation, it is again advantageous to program the factorization process as a separate sub-routine. This aspect will also be discussed in more detail later. In this factorization process there are only two aspects to determine: the elements in the left-hand factor matrix and the new values of the elements in the reduced coefficient matrix. These values can be computed using the Gauss elimination equations described in the bi-factorization method.

In the first reduction step, only three values need be calculated. These are:

$$l_{22} = \frac{1}{a_{22}} = \frac{1}{2} \qquad \text{(stored in place of } a_{22}\text{)}$$

$$l_{12} = -\frac{a_{12}}{a_{22}} = \frac{1}{2} \qquad \text{(stored in place of } a_{12}\text{)}$$

$$a'_{11} = a_{11} - \frac{a_{12}a_{21}}{a_{22}}$$

$$= a_{11} - \frac{(a_{12})^2}{a_{22}} = \frac{5}{2} \qquad \text{(stored in place of } a_{11}\text{)}$$

After this reduction step, the first left-hand factor matrix and the reduced coefficient matrix are:

$$\begin{bmatrix} 1 & \frac{1}{2} & \cdot & \cdot \\ \cdot & \frac{1}{2} & \cdot & \cdot \\ \cdot & \cdot & 1 & \cdot \\ \cdot & \cdot & \cdot & 1 \end{bmatrix} \qquad \begin{bmatrix} \frac{5}{2} & \cdot & -1 & \cdot \\ \cdot & 1 & \cdot & \cdot \\ -1 & \cdot & 2 & -1 \\ \cdot & \cdot & -1 & 1 \end{bmatrix}$$

left-hand matrix reduced matrix

The same steps can be achieved computationally as follows. From NORD, the first column to be eliminated is column 2, since NORD (2) = 1. This locates the required pivot, DIAG (2), and thus l_{22} can be computed. From ICAP (2), the location of the first non-zero off-diagonal element is determined, location 3, and therefore l_{12} can be computed. From NEXT (3) = 0, there are no other non-zero off-diagonal elements in column 2, and therefore the only element to be affected in the reduced matrix is that in row 1 and column 1, i.e. a_{11}. Therefore the new value a'_{11} can be computed. After this first reduction, the values of the elements stored in the two tables are:

Location	1	2	3	4	5	6*	7	8	9	10
VALUE	–	−1·0	0·5	–	−1·0	–	–	–	–	–
IROW	–	3	1	–	4	–	–	–	–	–
NEXT	7	0	0	1	0	4	8	9	10	–

Location	1	2	3	4
DIAG	2·5	0·5	2·0	1·0
ICAP	2	3	5	0
NOZE	1	1	1	0
NORD	2	1	3	4

Using the same principles for the numerical calculations at the second and third reduction steps gives the following values in the arrays VALUE and DIAG only:

Second reduction step:

Location	1	2	3	4	5	6	7	8	9	10
VALUE	–	0·4	0·5	–	–1·0	–	–	–	–	–
DIAG	0·4	0·5	1·6	1						

Third reduction step:

Location	1	2	3	4	5	6	7	8	9	10
VALUE	–	0·4	0·5	–	0·625	–	–	–	–	–
DIAG	0·4	0·5	0·625	0·375						

Similarly, after the final reduction step, the two tables are:

Location	1	2	3	4	5	* 6	7	8	9	10
VALUE	–	0·4	0·5	–	0·625	–	–	–	–	–
IROW	–	3	1	–	4	–	–	–	–	–
NEXT	7	0	0	1	0	4	8	9	10	–

Location	1	2	3	4
DIAG	0·4	0·5	0·625	2·6$\dot{6}$
ICAP	2	3	5	0
NOZE	1	1	1	0
NORD	2	1	3	4

With this information in store, all the factor matrices, both left and right, can be restructured, and hence, from the definition of the factor matrices, the solution of a set of equations **AX** = **b** can be computed. Also, it is evident that, if a new coefficient matrix having the same sparsity structure needs to be factorized, the values of the elements in VALUE and DIAG can be re-defined and factorization repeated using the same simulated ordering. Both of these aspects will be discussed in more detail in section 6.6.

6.5 Programming principles of transformation matrices

As described in chapter 4, elementary transformation matrices are created during factorization at each stage of elimination. This occurs whichever method of factorization is used, although the types of factor matrices will be different in each method. These matrices are fundamental in the direct solution of linear equations and therefore the programming should enable them to be stored and manipulated in subsequent calculations as efficiently as possible. The programming principles involved in the creation of the non-zero elements of these factor matrices were described in the previous section, but the subsequent manipulation of factor matrices was not considered. Therefore, to indicate some of the most pertinent points to consider when programming the storage and manipulation of transformation matrices, consider the two typical matrices shown below:

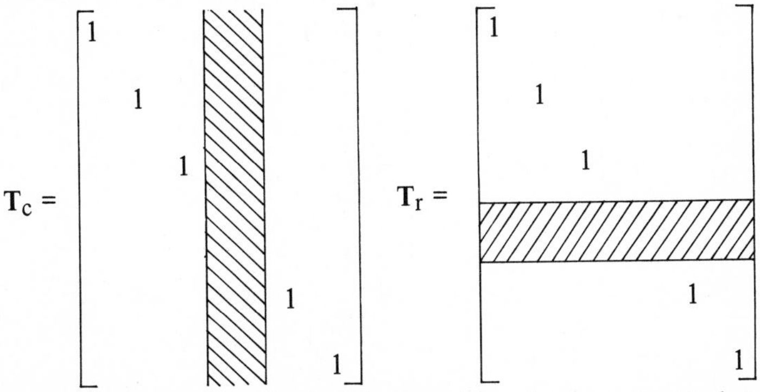

The reader will soon verify that these matrices have an identical structure to those created during all factorization methods. Consequently the following discussion relates to all types of factorization.

Consider first the various aspects concerning storage of this type of matrix. An important point to stress is that it is not essential to store the complete matrix. It is sufficient to store only the significant column (shaded area in T_c) and the significant row (shaded area in T_r), since the only other non-zero elements are diagonal elements, all of which are unity. Therefore in the subsequent calculations all diagonal elements other than those in the relevant column or row are assumed implicitly to be unity.

In the case of symmetrical coefficient matrices, which occur with most of the physical network type of problems, some methods, in particular bi-factorization, create factor matrices which, except for the diagonal elements, are the transposition of each other. Therefore it is sufficient to store only one of the matrices. Generally the left-hand factor matrix in the bi-factorization is stored, because its diagonal element is not unity.

The programming principles involved in the storage of the non-zero elements of the left-hand factor matrix produced during bi-factorization were also described in the previous section. The same principles can be applied to the storage of any transformation matrix. Since the transformation matrix may be sparse and only the non-zero elements need be stored, it is preferable that the storage scheme should include the linked list technique. This method offers a significant advantage, since the subsequent calculations involving transformation matrices are simple multiplications. These multiplications are carried out column-wise or row-wise and the linked list technique requires minimal searching and indexing.

Consider now the subsequent manipulation of these stored transformation matrices. Generally this manipulation involves multiplication of a vector by the transformation matrix to produce a new vector. This process may continue sequentially with further transformation matrices until the required vector is produced. This type of multiplication can be carried out very simply with the minimum number of arithmetical operations, as described below.

Consider the multiplication of a vector by a transformation matrix having one significant column, that is:

$$\begin{bmatrix} 1 & & a_{1i} & & \\ & \ddots & \vdots & & \\ & & 1 \quad \vdots & & \\ & & a_{ii} & & \\ & & \vdots & 1 & \\ & & \vdots & & \ddots \\ & & a_{ni} & & 1 \end{bmatrix} \begin{bmatrix} b_1 \\ \vdots \\ b_i \\ \vdots \\ b_n \end{bmatrix} = \begin{bmatrix} b'_1 \\ \vdots \\ b'_i \\ \vdots \\ b'_n \end{bmatrix}$$

$$\mathbf{T_i b} = \mathbf{b'}$$

The elements of vector $\mathbf{b'}$ are calculated from:

$$b'_j = b_j + a_{ji} b_i \qquad j = 1,2 \ldots, n \neq i$$

and $b'_i = a_{ii} b_i$

For zero values of a_{ji}, the values of b'_j are the same as b_j. Therefore, the above equation needs to be used only for the non-zero elements of the transformation matrix. It is now evident how a linked list method of storage assists in this manipulation. By commencing at the first non-zero element, the appropriate element of the vector is modified. The linked list indicates the next non-zero element, and the next appropriate element of the vector is modified. Consequently only those elements of the vector which need modification are affected; all others remain unchanged and, most important, are not detected or processed in any way.

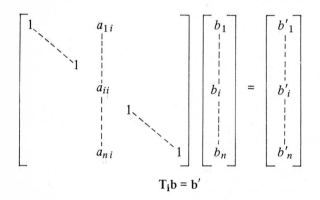

Consider now the multiplication of a vector by a transformation matrix having one significant row, that is:

$$\begin{bmatrix} 1 & & & & \\ & \ddots & & & \\ & & 1 & & \\ a_{i1} & \cdots & a_{ii} & \cdots & a_{in} \\ & & & 1 & \\ & & & & \ddots \\ & & & & & 1 \end{bmatrix} \begin{bmatrix} b_1 \\ \vdots \\ b_i \\ \vdots \\ b_n \end{bmatrix} = \begin{bmatrix} b'_1 \\ \vdots \\ b'_i \\ \vdots \\ b'_n \end{bmatrix}$$

$$\mathbf{T}_i\mathbf{b} = \mathbf{b}'$$

In this case, all the elements of $\mathbf{b}'$ are the same as those of $\mathbf{b}$, except for the element b'_i which is calculated from:

$$b'_i = \sum_{j=1}^{n} a_{ij}b_j$$

As in the previous case, the zero values of a_{ij} do not affect the appropriate element in the vector and therefore need not be considered. It is again evident that a linked list will assist in this computation very efficiently, since the only elements in the vector which will be processed will be those that must be multiplied by a non-zero value of a_{ij}.

Let us now consider the basic principles involved in using the information stored in the linked list scheme to solve a set of simultaneous linear equations. It can be advantageous to formulate this part of the overall program as a separate sub-routine, so that the equations can be solved for the same factorized coefficient matrix but with different right-hand vectors. This aspect will be discussed in more detail in the next section. To illustrate these principles, consider the solution of the equations:

$$\begin{bmatrix} 3 & -1 & -1 & \cdot \\ -1 & 2 & \cdot & \cdot \\ -1 & \cdot & 2 & -1 \\ \cdot & \cdot & -1 & 1 \end{bmatrix} \begin{bmatrix} x_1 \\ x_2 \\ x_3 \\ x_4 \end{bmatrix} = \begin{bmatrix} 1 \\ 1 \\ 1 \\ 1 \end{bmatrix}$$

that is, $\mathbf{AX} = \mathbf{b}$

Using the diagonal elements as pivots in the order 2, 1, 3 and 4 and the bi-factorization method gives the following solution to these equations:

$$\begin{bmatrix} x_1 \\ x_2 \\ x_3 \\ x_4 \end{bmatrix} = \underset{\mathbf{R}^{(1)}}{\begin{bmatrix} 1 & \cdot & \cdot & \cdot \\ \frac{1}{2} & 1 & \cdot & \cdot \\ \cdot & \cdot & 1 & \cdot \\ \cdot & \cdot & \cdot & 1 \end{bmatrix}} \underset{\mathbf{R}^{(2)}}{\begin{bmatrix} 1 & \cdot & \frac{2}{5} & \cdot \\ \cdot & 1 & \cdot & \cdot \\ \cdot & \cdot & 1 & \cdot \\ \cdot & \cdot & \cdot & 1 \end{bmatrix}} \underset{\mathbf{R}^{(3)}}{\begin{bmatrix} 1 & \cdot & \cdot & \cdot \\ \cdot & 1 & \cdot & \cdot \\ \cdot & \cdot & 1 & \frac{5}{8} \\ \cdot & \cdot & \cdot & 1 \end{bmatrix}} \underset{\mathbf{L}^{(4)}}{\begin{bmatrix} 1 & \cdot & \cdot & \cdot \\ \cdot & 1 & \cdot & \cdot \\ \cdot & \cdot & 1 & \cdot \\ \cdot & \cdot & \cdot & \frac{8}{3} \end{bmatrix}}$$

$$\underset{\mathbf{L}^{(3)}}{\begin{bmatrix} 1 & \cdot & \cdot & \cdot \\ \cdot & 1 & \cdot & \cdot \\ \cdot & \cdot & \frac{5}{8} & \cdot \\ \cdot & \cdot & \frac{5}{8} & 1 \end{bmatrix}} \underset{\mathbf{L}^{(2)}}{\begin{bmatrix} \frac{2}{5} & \cdot & \cdot & \cdot \\ \cdot & 1 & \cdot & \cdot \\ \frac{2}{5} & \cdot & 1 & \cdot \\ \cdot & \cdot & \cdot & 1 \end{bmatrix}} \underset{\mathbf{L}^{(1)}}{\begin{bmatrix} 1 & \frac{1}{2} & \cdot & \cdot \\ \cdot & \frac{1}{2} & \cdot & \cdot \\ \cdot & \cdot & 1 & \cdot \\ \cdot & \cdot & \cdot & 1 \end{bmatrix}} \underset{\mathbf{b}}{\begin{bmatrix} 1 \\ 1 \\ 1 \\ 1 \end{bmatrix}} = \underset{\mathbf{X}}{\begin{bmatrix} \frac{7}{3} \\ \frac{5}{3} \\ \frac{13}{3} \\ \frac{16}{3} \end{bmatrix}}$$

This same sequence of multiplication can be achieved computationally using the last set of tables given in section 6.4, as follows:

Consider the multiplication of $\mathbf{L}^{(1)}\mathbf{b}$.

From NORD, the first diagonal element chosen as pivot was a_{22}.

From DIAG (2), $\quad a_{22} = 0{\cdot}5$

$$\begin{aligned} \therefore \quad b'_2 &= a_{22}b_2 \\ &= 0{\cdot}5 \times 1 \\ &= 0{\cdot}5 \end{aligned}$$

From ICAP (2), the first non-zero off-diagonal element is in location 3 of the first table, and VALUE (3) = 0·5. From IROW (3), this element is in row 1, i.e. it is the element a_{12}.

$$\begin{aligned} \therefore \ b'_1 &= b_1 + a_{12}b_2 \\ &= 1 + (0{\cdot}5 \times 1) \\ &= 1{\cdot}5 \end{aligned}$$

From NEXT (3), there are no further non-zero off-diagonal elements, and therefore all other values of **b** remain unchanged.

This series of logical steps has shown that:

$$\mathbf{L}^{(1)}\mathbf{b} = \begin{bmatrix} 1 & 0{\cdot}5 & \cdot & \cdot \\ \cdot & 0{\cdot}5 & \cdot & \cdot \\ \cdot & \cdot & 1 & \cdot \\ \cdot & \cdot & \cdot & 1 \end{bmatrix} \begin{bmatrix} 1 \\ 1 \\ 1 \\ 1 \end{bmatrix} = \begin{bmatrix} 1{\cdot}5 \\ 0{\cdot}5 \\ 1 \\ 1 \end{bmatrix}$$

This process can be continued sequentially in the order given by NORD until all the left-hand factor matrices have been used, that is, $\mathbf{L}^{(1)}$, $\mathbf{L}^{(2)}$, $\mathbf{L}^{(3)}$ and $\mathbf{L}^{(4)}$. The same principles apply when multiplying by $\mathbf{R}^{(3)}$, $\mathbf{R}^{(2)}$ and $\mathbf{R}^{(1)}$, but as these factors are not stored explicitly it is perhaps useful to indicate the changes in identification. It is unnecessary to multiply by $\mathbf{R}^{(4)}$ because this is a unit matrix.

To illustrate these points, let the multiplication of:

$$\mathbf{L}^{(4)}\,\mathbf{L}^{(3)}\,\mathbf{L}^{(2)}\,\mathbf{L}^{(1)}\,\mathbf{b} = \mathbf{b}'$$

and consider the multiplication of:

$$\mathbf{R}^{(3)}\,\mathbf{b}' = \mathbf{b}''$$

which, in matrix form is:

$$\underset{\mathbf{R}^{(3)}}{\begin{bmatrix} 1 & \cdot & \cdot & \cdot \\ \cdot & 1 & \cdot & \cdot \\ \cdot & \cdot & 1 & \frac{5}{8} \\ \cdot & \cdot & \cdot & 1 \end{bmatrix}} \underset{\mathbf{b}'}{\begin{bmatrix} \frac{3}{5} \\ \frac{1}{2} \\ 1 \\ \frac{16}{3} \end{bmatrix}} = \underset{\mathbf{b}''}{\begin{bmatrix} \frac{3}{5} \\ \frac{1}{2} \\ \frac{13}{3} \\ \frac{16}{3} \end{bmatrix}}$$

This process can be achieved computationally as follows: The elements to be considered are those indexed by NORD (3). From ICAP (3), the first of these elements are given in location 5 of the first table, and from NEXT (5), only one non-zero off-diagonal element exists. The only significant difference in interpreting the position of this element to that discussed above is that IROW (5) = 4 indicates the element to be in the fourth column instead of the fourth row, that is, the element in question is a_{34}, not a_{43}. Therefore, from VALUE (5),

$$a_{34} = 0{\cdot}625$$

and $b''_3 = a_{33}b'_3 + a_{34}b'_4$

Now all the diagonal elements of the right-hand factor matrices are unity,

$$\begin{aligned} \therefore\ b''_3 &= b'_3 + a_{34}b'_4 \\ &= 1 + (0{\cdot}625 \times 5{\cdot}3333) \\ &= 4{\cdot}3333 = \frac{13}{3} \end{aligned}$$

This process can then be continued in the reverse order given by NORD until all the right-hand factor matrices have been used and the solution, as given above, is obtained.

6.6 Principles of program organization

At this point it should be evident to a reader that several distinct aspects are involved in the analysis of large, sparse network-type problems; the equations describing the problem must be formulated, the sequence of elimination established, the coefficient matrix factorized and, finally, the problem numerically solved. These four aspects can be included in one overall program or treated and programmed as separate sub-routines. Separation of the various aspects into distinct sub-routines offers a very significant advantage.

Generally, the problems that arise in practice can be categorized into three main groups. These are:

(i) a single solution is required for a given coefficient matrix and a given right-hand vector;

(ii) several solutions are required for a given coefficient matrix but with different right-hand vectors;

(iii) several solutions are required for different coefficient matrices which have the same incidence structure but different numerical values.

It is evident from the previous sections of this chapter that sparsity programming can be relatively complicated. It should therefore be done efficiently and the program should be organized in such a way that solutions to the different groups of problems can be achieved without unnecessary repetition of calculations. For example, when the solution is required for several right-hand vectors, the same factorized matrix can be used. Similarly, when several solutions are required with different coefficient matrices having the same structure, the original pivoting sequence can be used for all the required solutions.

From this discussion it should be evident that the overall problem is conveniently subdivided into its constituent parts, i.e. formulation, ordering, factorization and solution, as discussed above. The various programming aspects of these different sub-routines have already been discussed. As discussed later, this subdivision of the program achieves considerable flexibility. The basic principles involved in these four sub-routines are as follows:

(i) Compacting – This involves the generation of the non-zero coefficients of the equations from the basic information for the network and the storing of them in a compact form. Since only the non-zero coefficients are stored and not the complete matrix, it is necessary to incorporate an efficient addressing scheme to locate and identify each element, whilst minimizing the additional storage required.

(ii) Simulation and Ordering – This establishes the order of elimination. With dynamic ordering it is preferable to simulate the effect of sparsity on the elimination before commencing actual floating point arithmetic. This simulation process can make the extra time spent in ordering very worthwhile, since once the order has been established, it will remain unchanged for any problem having the same sparsity structure. For instance, there are many problems which require iterative solution. Although the coefficients may change numerically from one iteration to another, the sparsity structure can remain the same. An example of this occurs in the solution of non-linear equations using successive linearization. For problems such as this, it is necessary only to factorize the matrix from iteration to iteration using the order of elimination obtained from the initial simulation.

(iii) Factorization – This factorizes the coefficient matrix into any one of the factored forms as discussed in chapter 4, using the order of elimination deduced during the simulation process. The operations are performed only on the non-zero elements and only the non-zero elements of the factorized matrix are stored. For symmetrical coefficient matrices using bi-factorization or a suitably modified triangular decomposition method only the non-zero elements of the left-hand or lower triangular matrix need be stored, which reduces the required storage to about one-half.

(iv) Numerical Solution – This is the final process of the calculations and involves computing the numerical values of the unknown variables using the factor matrices and the known vector of the original equations. By separating this calculation from those discussed above, it is possible to solve the same basic equations for various known vectors very easily.

Using this sequence of sub-routines gives simplified main program flow charts as shown in Fig. 6.2. The three alternative schemes show clearly the advantage gained in solving the different groups of problems discussed previously with the one main program.

6.7 Computational analysis of typical systems

The authors have extensively used a program based on the four sub-routines and the programming principles discussed in this chapter, using bi-factorization and dynamic ordering. Due to the physical nature of the problems with which they have been concerned, the diagonal elements of the associated coefficient matrices have been numerically dominant. This, together with the great

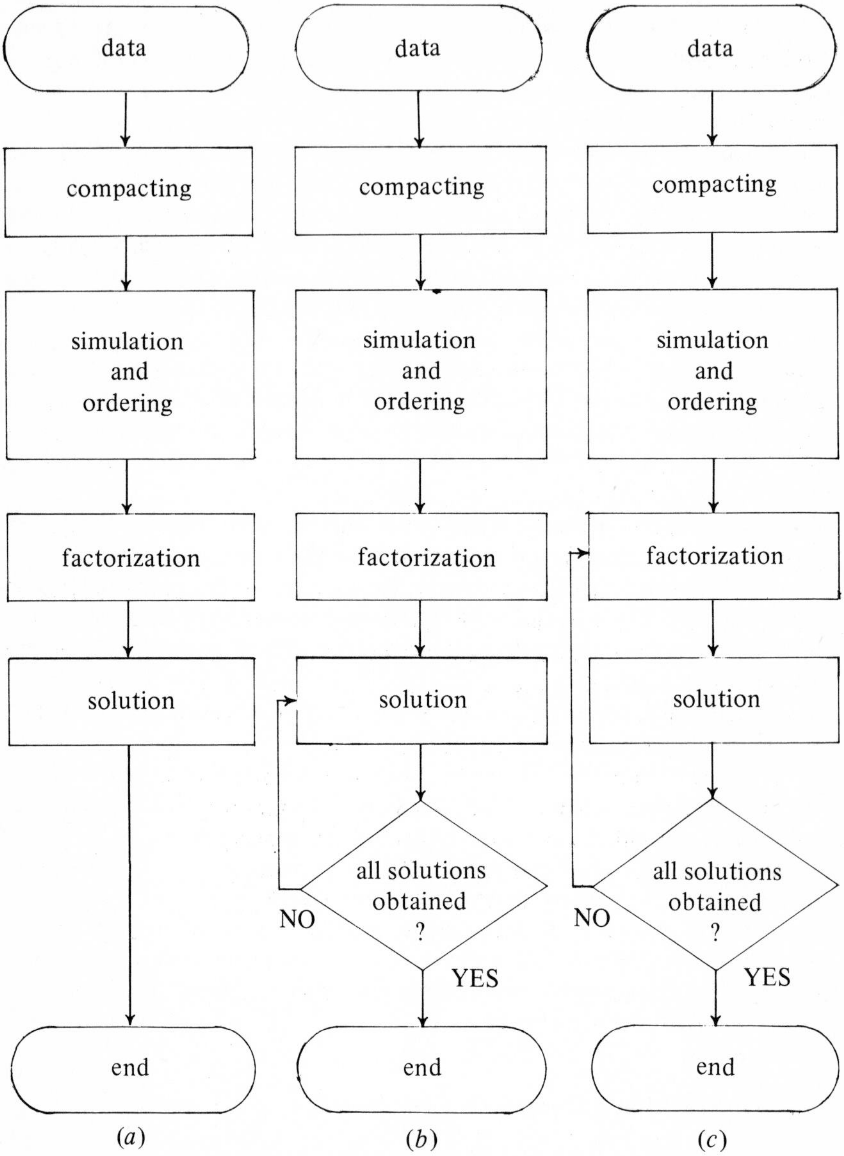

Fig. 6.2 Typical main program flow charts for analysis of network problems. (*a*) single solution; (*b*) multiple solution with different right-hand vectors; (*c*) multiple solution with different coefficient matrices having same structure accuracy of modern computers, has allowed them to choose, quite safely, pivots from the diagonal elements only. This increases computational efficiency, since no other pivoting has been required. In cases where the diagonal elements tend to vanish relatively or are numerically smaller than off-diagonal elements, partial pivoting may become necessary to improve the overall accuracy at the expense of computational time.

To illustrate the efficiency of sparsity-directed analysis of large, sparse networks, the number of non-zero elements of the original matrix, the number of non-zero elements of the factor matrices needed to be stored and the computational times of each sub-routine are shown in Table 6.1 for different types of networks and for different numbers of nodes and branches. These results were obtained with a CDC 7600 computer.

These networks illustrate the computational effort for two extreme types of problems: the random type of network and the regular type of network. In the examples quoted, the random type of network is illustrated by typical electrical- or gas-distribution systems. These practical problems require no additional explanation. The regular network is typical of a numerical approximation method for Laplace's equation, in which the field or space is divided into regular meshes, as shown in Fig. 6.3.

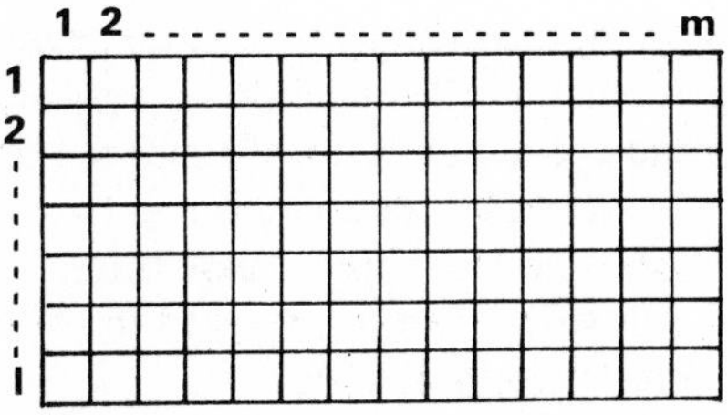

Fig. 6.3 Representation of Laplace equations

The intersections of these meshes are equivalent to nodes and the connections between the nodes are equivalent to branches of a network. This type of problem gives a coefficient matrix A of the block tridiagonal form:

$$A = \begin{bmatrix} T & -U & & & & \\ -U & T & -U & & & \\ & -U & T & & & \\ & & & \ddots & & \\ & & & & T & -U \\ & & & & -U & T \end{bmatrix}$$

$$\text{where } \mathbf{T} = \begin{bmatrix} 4 & -1 & & \cdots & & \\ -1 & 4 & -1 & \cdots & & \\ & -1 & 4 & \cdots & & \\ \vdots & \vdots & \vdots & \ddots & \vdots & \vdots \\ & & & \cdots & 4 & -1 \\ & & & \cdots & -1 & 4 \end{bmatrix}$$

A is of order n (= $m \times l$) and has l diagonal blocks,
U is a unit matrix of order m
and **T** is of order m.

The number of non-zero elements after elimination, quoted in Table 6.1, are those of the left-hand factor matrix only, since all the problems have symmetrical coefficient matrices and therefore only one factor matrix need be stored. It is therefore evident that in all the problems a certain degree of fill-in occurs, although for some types of networks, this fill-in is very small.

It is seen that the degree of fill-in and the computational times for the regular network type of problem are greater than for a random network having approximately the same number of nodes. This increase is inherent with this type of problem. However, it is still very advantageous to use sparsity techniques and programming for these problems.

The advantage of using the four sub-routines discussed previously is also evident from Table 6.1. Most of the computing time is spent in simulating the ordering process; with the large systems this component is by far the greatest. Consequently it is clear that if similar problems are to be solved, then once the order has been predetermined, the time taken to recalculate the problem for new coefficient matrices or new known vectors becomes minimal, even for the very large systems.

Another aspect which is made evident from the results in Table 6.1 is the tremendous advantage of the factorization method compared with matrix inversion. Considering the 1242 node power system type of problem, the final number of non-zero elements needed to be stored is about 4400. This compares with about $1\frac{1}{2}$ million if direct inversion techniques were used and indicates that the fill-in is relatively minute. Also, the combined elimination and solution time is about 200 ms and the ordering time is about $1\frac{1}{2}$ s. This indicates that little could be gained from the extra effort to achieve optimal ordering, since the ordering time would have been increased very significantly with little hope of any comparable reduction in either the number of stored non-zero elements or the combined factorization and solution time.

From this discussion and typical results it is hoped that the reader will

Table 6.1 – Storage and computation times for typical networks using bi-factorization, dynamic ordering and a CDC 7600 computer

network type	mesh divisions $m \times l$	number of		ratio $\frac{b}{n}$	Number of non-zero elements		computation times (s)			
		nodes (n)	branches (b)		initial	final	compacting	ordering	factorization	solution
P	–	14	20	1·43	54	38	0·001	0·001	0·001	0·0005
	–	30	41	1·37	112	85	0·001	0·003	0·003	0·0007
	–	57	80	1·40	217	200	0·003	0·007	0·007	0·001
	–	118	184	1·56	486	390	0·007	0·020	0·013	0·003
	–	621	794	1·28	2209	2136	0·030	0·422	0·082	0·017
	–	1242	1593	1·28	4428	4430	0·058	1·578	0·182	0·038
L	5 x 6	30	49	1·63	128	131	0·002	0·006	0·006	0·001
	10 x 11	110	199	1·81	508	715	0·010	0·050	0·047	0·006
	15 x 16	240	449	1·87	1138	1919	0·023	0·213	0·074	0·040
	30 x 31	930	1799	1·93	4528	10937	0·091	2·578	1·896	0·081

P = typical electrical- or gas-distribution networks, L = Laplacian or regular mesh problem divided into $m \times l$ meshes.

appreciate the tremendous advantages that can be gained in terms of computational efficiency if suitable programming is used to incorporate the effect of sparsity. These advantages are likely to become more important as time progresses and the users' demand of a given computer increases. Nowadays engineers and non-engineers alike are much more aware of cost-effectiveness and the problems of overcapitalization. Consequently utilization efficiency is of paramount importance, and the exploitation of sparsity in network analysis enables this efficiency to be very much increased.

Bibliography

Berry, R. D.: An optimal ordering of electronic circuit equations for a sparse matrix solution. *Inst. elect. electron. Engrs Trans.* (CT–18), **1971**, 40–50.

Branin, F. H.: Computer methods of network analysis. *Proc. Inst. elect. electron. Engrs* **55**, 1787–1801, 1967.

Buchet, J. de: How to take into account the low density of matrices to design a mathematical programming package. In *Large Sparse Sets of Linear Equations,* pp. 211–18. Academic Press, 1971.

Churchill, M. E.: A sparse matrix procedure for power system analysis programs. In *Large Sparse Sets of Linear Equations,* pp. 127–38. Academic Press, 1971.

Curtis, A. R. & Reid, J. K.: *Fortran Sub-routines for the Solution of Sparse Sets of Linear Equations* (Report R.6844). United Kingdom Atomic Energy Authority, 1971.

Curtis, A. R. & Reid, J. K.: The solution of large sparse systems of linear equations. *Proc. Int. Fedn Inf. Processg* (Report TP 450). United Kingdom Atomic Energy Authority, 1971.

Gustavson, F. G.: Some basic techniques for solving sparse systems of linear equations. In *Sparse Matrices and Their Applications,* pp. 41–52. Plenum Press, 1972.

Gustavson, F. G., Liniger, W. & Willoughby, R. A.: Symbolic generation of an optimal Crout algorithm for sparse systems of equations. *Ass. Computg Mach. J.* **17**, 87–109, 1970.

Hachtel, G., Brayton, R. & Gustavson, F.: The sparse tableau approach to network analysis and design. *Inst. elect. electron. Engrs Trans.* (CT–18), **1971**, 101–13.

Jennings, A.: A compact storage scheme for the solution of symmetrical linear simultaneous equations. *Comput. J.* **9**, 281–5, 1966.

Jennings, A.: A sparse matrix scheme for the computer analysis of structures. *Int. Jl Comput. Math.* **2**, 1–21, 1968.

Jennings, A. & Tuff, A. D.: A direct method for the solution of large sparse symmetric simultaneous equations. In *Large Sparse Sets of Linear Equations,* pp. 97–104. Academic Press, 1971.

Jensen, H. G.: Efficient matrix techniques applied to transmission tower design. *Proc. Inst. elect. electron. Engrs* **55**, 1997–2000, 1967.

Jensen, H. G. & Parks, G. A.: Efficient solutions for linear matrix equations. *J. struct. Div., Proc. Am. Soc. civ. Engrs* **96**, 49–64, 1970.

Jodeit, J. G.: Storage organisation in programming systems. *Commun. Ass. Computg Mach.* **11**, 741–6, 1968.

Kettler, P. C. & Weil, R. L.: An algorithm to provide structure for decomposition. In *Sparse Matrix Proceedings,* pp. 11–24. IBM, 1969.

Larcombe, M. H. E.: A list processing approach to the solution of large sets of matrix equations and the factorisation of the overall matrix. In *Large Sparse Sets of Linear Equations,* pp. 25–40. Academic Press, 1971.

Lee, H. B.: An implementation of Gaussian elimination for sparse systems of linear equations. In *Sparse Matrix Proceedings,* pp. 75–84. IBM, 1969.

Maurser, W. D.: *Programming, Introduction to Computer Languages and Techniques.* Holden-Day, 1968.
Nuding, E. & Kahlert-Warmbold, I.: A computer oriented representation of matrices. *Computing* **6**, 1–8, 1970.
Ogbuobiri, E. C.: Dynamic storage and retrieval in sparsity programming. *Inst. elect. electron. Engrs Trans.* (PAS–89), **1970**, 150–5.
Randell, B. & Kuehner, C. J.: Dynamic storage allocation systems. *Commun. Ass. Computg Mach.* **11**, 279–306, 1968.
Tewarson, R. P.: Computations with sparse matrices. *Symp. appl. Math. Rev.* **12**, 527–44, 1970.
Willoughby, R. A.: Sparse matrix algorithms and their relation to problem classes and computer architecture. In *Large Sparse Sets of Linear Equations,* pp. 255–77. Academic Press, 1971.
Zollenkopf, K.: Bi-factorisation: basic computational algorithm and programming techniques. In *Large Sparse Sets of Linear Equations,* pp. 75–96. Academic Press, 1971.

7 Application of Sparsity Techniques to Linear Programming Problems

7.1 Introduction

The sparsity techniques discussed in previous chapters can be applied to many engineering and non-engineering problems. The most obvious is that of a physical linear network problem for which all the equations are of the form $\mathbf{AX} = \mathbf{b}$ and the number of equations is equal to the number of unknowns. In these cases the sparsity techniques can be applied directly in the form described earlier and illustrated by the various numerical examples quoted in the text. The reader can therefore be left to apply these techniques to his own particular problems.

There are, however, several problems for which the advantageous uses of sparsity techniques are not so obvious or direct. Two examples of this concern linear programming problems and non-linear network problems which are solved by successive linearization at each iterative step. This chapter is devoted to some aspects of linear programming problems in order to indicate how sparsity-directed programming can be beneficially included in their solution. The next chapter considers non-linear network problems.

Linear programming is concerned with the general class of optimization problems that are defined by a linear expression subject to a number of linear constraints. In practice it is sometimes necessary to solve these linear programming problems when the number of constraints is very large and when they are sparse. Sparsity methods and programming can be applied advantageously to such problems.

There are many types of linear programming problems to which sparsity techniques can be applied and it is outside the scope of this book to cover all of them. To show how the technique developed in the preceding chapters can be applied, however, two important aspects of linear programming are considered. The first is the revised simplex method using the product form of the inverse matrix and the second is the application of sparsity to the transshipment problem.

This book is not a text on linear programming and therefore it is assumed that the reader is familiar with the mathematical concepts and techniques of linear programming. Some of these aspects will be dealt with, however, in order to develop the application of the sparsity techniques. Also, although only two aspects of linear programming are covered, they are described in sufficient detail to be of direct practical use and to allow the reader to apply the same techniques to other linear programming aspects as necessary.

7.2 Application to the revised simplex method

7.2.1 Basic feasible solution

The revised simplex method, developed by Dantzig, is widely used to solve large linear programming problems on a digital computer. It is expected that the reader will be familiar with the fundamental aspects of this method and only a summary of the essential details is given in this chapter. If the reader wishes to understand more, there are several excellent texts on the subject.

In general, the essence of a linear programming problem can be defined as:

$$\min \quad Z = \mathbf{CX} \tag{7.1}$$

subject to

$$\mathbf{AX} = \mathbf{b} \tag{7.2}$$

and

$$\mathbf{X} \geqslant 0 \tag{7.3}$$

where

Z = objective function

$\mathbf{C}$ = cost coefficients

$\mathbf{X}$ = unknown variables including the slack and surplus quantities

$\mathbf{A}$ = coefficient matrix of constraints

$\mathbf{b}$ = known right-hand vector

The optimization problem defined by equation 7.1 may involve maximization of the objective function and not minimization as given above. These problems can be formulated as minimization problems by changing the sign of the objective function. Therefore only the minimization problem as defined by equation 7.1 will be considered in this chapter.

At each iterative step of the solution of a linear programming problem, matrix $\mathbf{A}$ can be partitioned to give a square non-singular matrix $\mathbf{B}$ and a set of corresponding variables $\mathbf{X}_b$. Partitioning equation 7.2 in this way gives:

$$[\mathbf{A}' \mid \mathbf{B}] \begin{bmatrix} \mathbf{X}_a \\ \hline \mathbf{X}_b \end{bmatrix} = [\mathbf{b}] \tag{7.4}$$

Expanding equation 7.4 gives:

$$\mathbf{A}'\mathbf{X}_a + \mathbf{BX}_b = \mathbf{b} \tag{7.5}$$

By partitioning the cost function in a similar manner, equation 7.1 becomes:

$$Z = \mathbf{C}_a \mathbf{X}_a + \mathbf{C}_b \mathbf{X}_b \tag{7.6}$$

Because **B** is square and non-singular, the matrix can be inverted. Therefore from equation 7.5, $\mathbf{X}_b$ can be expressed as:

$$\mathbf{X}_b = \mathbf{B}^{-1} (\mathbf{b} - \mathbf{A}'\mathbf{X}_a) \tag{7.7}$$

In order to satisfy the constraints given by equation 7.3, the elements of $\mathbf{B}^{-1}\mathbf{b}$ must be positive. Also, if the variables $\mathbf{X}_a$ are made zero, equation 7.7 becomes:

$$\mathbf{X}_b = \mathbf{B}^{-1}\mathbf{b} \tag{7.8}$$

Since equation 7.8 satisfies all the constraints of the minimization problem, the solution given by this equation is a possible solution to the problem. It is defined as a *basic feasible* solution. Furthermore, the variables $\mathbf{X}_b$ are defined as the *basic* variables, the variables $\mathbf{X}_a$ as the *non-basic* variables and the matrix **B** as a *basis* matrix.

Substituting for $\mathbf{X}_b$ from equation 7.7 into equation 7.6 gives:

$$Z = \mathbf{C}_b \mathbf{B}^{-1} \mathbf{b} + \mathbf{C}_a' \mathbf{X}_a \tag{7.9}$$

where $$\mathbf{C}_a' = \mathbf{C}_a - \mathbf{C}_b \mathbf{B}^{-1} \mathbf{A}' \tag{7.10}$$

7.2.2 Improving a basic feasible solution

Although the basic feasible solution given by equation 7.8 satisfies the constraints given by equations 7.2 and 7.3, it may not be the optimum solution required by equation 7.1. The value of the objective function specified by equation 7.1 depends on the values of the variables $\mathbf{X}_a$ (see equation 7.9). The basic feasible solution was derived, however, by assuming and defining the non-basic variables $\mathbf{X}_a$ to be zero. It is possible that, if some of these variables had a value greater than zero (equation 7.3 prevents them being negative), the objective function could decrease in value as required.

Suppose, therefore, that a non-basic variable x_i is increased in value. If the corresponding coefficient c_i' is negative, the value of the objective function as given by equation 7.9 will decrease, indicating that the basic feasible solution previously derived is not the optimum. The revised simplex method enables the variable x_i, corresponding to the most negative coefficient c_i', to be chosen and changed to a basic variable. This process is continued iteratively until all the coefficients c_i' are either positive or zero. In this case the objective function cannot be decreased further and the most recent basic feasible solution is the optimal solution.

7.2.3 Interchanging basic and non-basic variables

If from the process described in section 7.2.2 it is found that the objective function would be decreased by changing a non-basic variable x_i to a basic variable, one of the existing basic variables must now be changed to a non-basic variable. The problem is to find which basic variable must be brought out of the basis.

Consider equation 7.7, which can be expressed as:

$$\mathbf{X}_b = \mathbf{b}' - \mathbf{B}^{-1}\mathbf{A}'\mathbf{X}_a \tag{7.11}$$

where $\mathbf{b}' = \mathbf{B}^{-1}\mathbf{b}$ (7.12)

Since only one non-basic variable x_i is to be increased from zero, equation 7.11 can be written in terms of this non-basic variable only, that is:

$$\mathbf{X}_b = \mathbf{b}' - \mathbf{B}^{-1}\mathbf{P}_i x_i \tag{7.13}$$

where $\mathbf{P}_i$ is column i of matrix $\mathbf{A}'$.

Equation 7.13 can be expressed as:

$$\mathbf{X}_b = \mathbf{b}' - \mathbf{P}_i' x_i \tag{7.14}$$

where $\mathbf{P}_i' = \mathbf{B}^{-1}\mathbf{P}_i$ (7.15)

When the variable x_i is introduced into the basis, the existing basic variable which is removed must be reduced to a value of zero without making any of the other variables negative. To determine which variable to remove and satisfy this criterion, an index j is found which corresponds to the j-th row of the basis and which satisfies the conditions:

$$\min \frac{b_j'}{p_{ji}'} \tag{7.16}$$

and $p_{ji}' > 0$ (7.17)

where b_j' is the j-th element of $\mathbf{b}'$

and p_{ji}' is the j-th element of $\mathbf{P}_i'$

The variable which satisfies the conditions specified by equations 7.16 and 7.17 is made a non-basic variable and its value is reduced to zero.

At the r-th iterative step, the inverse of the new basis matrix $\mathbf{B}_r^{-1}$ can be evaluated from the inverse of the previous basis matrix $\mathbf{B}_{r-1}^{-1}$ using a transformation matrix $\mathbf{T}_r$, as described in chapter 4. Therefore:

$$\mathbf{B}_r^{-1} = \mathbf{T}_r\mathbf{B}_{r-1}^{-1} \tag{7.18}$$

where

$$\mathbf{T}_r = \begin{bmatrix} 1 & & & -\frac{p_{1i}'}{p_{ji}'} & & & \\ & \ddots & & \vdots & & & \\ & & 1 & \vdots & & & \\ & & & \frac{1}{p_{ji}'} & & & \\ & & & \vdots & 1 & & \\ & & & \vdots & & \ddots & \\ & & & -\frac{p_{mi}'}{p_{ji}'} & & & 1 \end{bmatrix} \tag{7.19}$$

and m = number of constraints.

By using artificial variables if necessary, it is always possible to start the iterative process with a unity basis matrix. At the end of the iterations these artificial variables will become non-basic variables, leaving only true, surplus or slack variables as basic variables. Using this technique, the inverse of the basis at any iterative step can be expressed solely as the product of the transformation matrices, that is, it is expressed in the product form of the inverse. The inverse of the r-th basis is therefore:

$$\mathbf{B}_r^{-1} = \mathbf{T}_r \mathbf{T}_{r-1} \ldots \mathbf{T}_2 \mathbf{T}_1 \tag{7.20}$$

If the original coefficient matrix of constraints is sparse, the significant column of these transformation matrices is likely also to be sparse. It is evident therefore that the sparsity techniques and programming discussed in previous chapters can be exploited in such cases, that is, only the non-zero elements of the significant column of these transformation matrices need be stored and processed.

7.2.4 Steps of solution

The following logical steps are used to program and compute the values of the basic variables which satisfy the optimization problem and constraints:

(i) Find the initial basic feasible solution. The iterative process is most conveniently commenced with a unity basis matrix, as discussed in the previous section.

(ii) At each iterative step calculate the elements of $\mathbf{C}'_{a(r)}$, where $\mathbf{C}'_{a(r)}$ represents $\mathbf{C}'_a$ at the r-th iteration. These can be found by substituting equation 7.20 into equation 7.10 to give:

$$\mathbf{C}'_{a(r)} = \mathbf{C}_{a(r)} - \mathbf{C}_{b(r)} \mathbf{T}_r \ldots \mathbf{T}_2 \mathbf{T}_1 \mathbf{A}'_{(r)} \tag{7.21}$$

(iii) If all the elements of $\mathbf{C}'_{a(r)}$ are greater than or equal to zero, the optimal solution has been found and the iterative process can be terminated. Otherwise find the non-basic variable x_i corresponding to the most negative element of $\mathbf{C}'_{a(r)}$ and enter this variable into the basis.

(iv) Find $\mathbf{P}'_{i(r)}$ from the equation obtained by substituting equation 7.20 into equation 7.15, that is, from:

$$\mathbf{P}'_{i(r)} = \mathbf{T}_r \ldots \mathbf{T}_2 \mathbf{T}_1 \mathbf{P}_{i(r)} \tag{7.22}$$

Find the basic variable x_j which must leave the basis from equations 7.16 and 7.17.

(v) Find the new transformation matrix $\mathbf{T}_{r+1}$, up-date $\mathbf{b}'$ and repeat steps (ii) to (v) until the optimal solution is found.

7.2.5 Storage scheme

Two matrices must be stored in the revised simplex method: the initial coefficient matrix $\mathbf{A}$ and the inverse of the basis in the product form, which must be up-dated at each iterative step.

It was shown in Chapter 6 how transformation matrices can be stored using sparsity programming so that the non-zero elements of the factor matrices use the storage positions of the original coefficient matrix. In the revised simplex method the original matrix is also required and therefore the elements of the product form of the inverse must be stored separately. This simplifies the storage scheme at the expense of increased storage. The scheme can be best illustrated by considering the following numerical example:

$$\mathbf{C} = \begin{array}{ccccccc} 1 & 2 & 3 & 4 & 5 & 6 & 7 \\ [-16, & -15, & 0, & 0, & 0, & 0, & 0] \end{array}$$

$$\mathbf{A} = \begin{array}{c} 4 \\ 5 \\ 6 \\ 7 \end{array} \left[\begin{array}{c|c|c|c|c|c|c} x_1 & x_2 & x_3 & x_4 & x_5 & x_6 & x_7 \\ 1 & \cdot & \cdot & 1 & \cdot & \cdot & \cdot \\ \cdot & 1 & \cdot & \cdot & 1 & \cdot & \cdot \\ 1 & \cdot & -1 & \cdot & \cdot & 1 & \cdot \\ \cdot & 1 & 1 & \cdot & \cdot & \cdot & 1 \end{array}\right]$$

$$\mathbf{b} = \begin{array}{c} 4 \\ 5 \\ 6 \\ 7 \end{array} \begin{bmatrix} 20{\cdot}0 \\ 20{\cdot}0 \\ 15{\cdot}0 \\ 15{\cdot}0 \end{bmatrix} \qquad \mathbf{X}_b = \begin{array}{c} 4 \\ 5 \\ 6 \\ 7 \end{array} \begin{bmatrix} x_4 \\ x_5 \\ x_6 \\ x_7 \end{bmatrix}$$

The indices (4–7) used in conjunction with the above matrix and vectors indicate the variables which are in the basis. These therefore serve as a means of identification.

The coefficient matrix **A** can be stored in a compact form using the storage techniques described in chapter 6. In this case only four arrays arranged in two tables are necessary, these being VALUEA (numerical value of element in matrix **A**), IROWA (index of row), ICAPA (index of column address pointer) and NOZEA (number of non-zero elements in each column of **A**). The two tables for the above example are therefore:

Location	1	2	3	4	5	6	7	8	9	10
VALUEA	1·0	1·0	1·0	1·0	−1·0	1·0	1·0	1·0	1·0	1·0
IROWA	1	3	2	4	3	4	1	2	3	4

Location	1	2	3	4	5	6	7
ICAPA	1	3	5	7	8	9	10
NOZEA	2	2	2	1	1	1	1

Any column of matrix **A** can be reconstructed very simply. Consider the reconstruction of column 3. From the second table, the number of non-zero elements in column 3 is given by:

NOZEA (3) = 2

The location of the first element in column 3 is given by:

ICAPA (3) = 5

Therefore the non-zero elements of column 3 are given in locations 5 and 6 of the first table. This table indicates that the values of these elements and their corresponding row positions are:

VALUEA (5) = −1·0 and IROWA (5) = 3

VALUEA (6) = 1·0 and IROWA (6) = 4

A similar scheme can be used to store the transformation matrices **T** constituting the product form of the inverse of the basis matrix. As discussed in section 7.2.3, only the significant column of these transformation matrices need be stored: the other diagonal elements are always equal to unity and can be assumed implicitly.

The numerical values and row positions of the non-zero elements of **T** are stored in a similar manner to those of **A**, using equivalent arrays VALUET and IROWT. Similarly the index of column address pointer and the number of non-zero elements in each significant column are stored in ICAPT and NOZET. In addition to these values, the position of the significant column must also be identified. The order in which these significant columns are created is irregular, since it depends upon which variables are interchanged during iteration. The most logical storage method is to store the columns consecutively when created and identify the column number in an additional array NCOLT attached to the second table.

7.2.6 Numerical example

To illustrate the steps of solution and the storage scheme given in the previous sections, consider the minimization problem defined by equations 7.1–7.3 for the values of **C**, **A** and **b** given in section 7.2.5.

Iteration 1

(i) As discussed previously, it is convenient to start with basic variables $\mathbf{X}_b$, which correspond to a unity basis matrix. In this example, therefore, the basic variables will be:

$$\mathbf{X}_b^T = [x_4, x_5, x_6, x_7]$$

and the corresponding values of $\mathbf{C}_b$ are:

$$\begin{matrix} & 4 & 5 & 6 & 7 \\ \mathbf{C}_b = [& 0, & 0, & 0, & 0] \end{matrix}$$

(ii) Since the initial basis matrix **B** is unity, the value of $\mathbf{C}_b\mathbf{B}^{-1}$ is the same as $\mathbf{C}_b$. Therefore, from equation 7.10,

$$\mathbf{C}'_a = \mathbf{C}_a - \mathbf{C}_b\mathbf{B}^{-1}\mathbf{A}'$$

$$\begin{array}{ccccccc} 1 & 2 & 3 & 4 & 5 & 6 & 7 \end{array}$$

$$= \left[\,\boxed{-16},\ -15,\ 0,\ *,\ *,\ *,\ *\,\right]$$

where the asterisks correspond to the variables that are in the basis.

(iii) The variable corresponding to the most negative coefficient (−16 as encircled above) is x_1, and therefore this variable is selected to go into the basis.

(iv) The column of matrix **A**′ corresponding to the variable x_1 is:

$$\mathbf{P}_1 = \begin{array}{c} 4 \\ 5 \\ 6 \\ 7 \end{array}\left[\begin{array}{c} 1 \\ 0 \\ 1 \\ 0 \end{array}\right]$$

Hence, from equations 7.15 and 7.22,

$$\mathbf{P}'_1 = \mathbf{B}^{-1}\mathbf{P}_1 = \mathbf{P}_1 = \begin{array}{c} 4 \\ 5 \\ 6 \\ 7 \end{array}\left[\begin{array}{c} 1 \\ 0 \\ \boxed{1} \\ 0 \end{array}\right]$$

From equation 7.17, x_4 and x_6 must be considered as possible variables to go outside the basis. Of these two variables, x_6 satisfies equation 7.16 and is therefore changed to a non-basic variable.

(v) The transformation matrix $\mathbf{T}_1$ is obtained from equation 7.19 as:

$$\begin{array}{cccc} 4 & 5 & 6 & 7 \end{array}$$

$$\mathbf{T}_1 = \begin{array}{c} 4 \\ 5 \\ 1 \\ 7 \end{array}\left[\begin{array}{cc|c|c} 1 & \cdot & -1 & \cdot \\ \cdot & 1 & \cdot & \cdot \\ \cdot & \cdot & 1 & \cdot \\ \cdot & \cdot & \cdot & 1 \end{array}\right]$$

As discussed in section 7.2.5, the significant column of this matrix is stored as follows:

Location	1	2	3	4	5	6	7	8
VALUET	−1·0	1·0	–	–	–	–	–	–
IROWT	1	3	–	–	–	–	–	–

Location	1	2	3	4
ICAPT	1	–	–	–
NOZET	2	–	–	–
NCOLT	3	–	–	–

From these two tables, ICAPT (1) = 1 indicates that the first non-zero element of the significant column of the first transformation matrix is stored in the first location of array VALUET, NOZET (1) = 2 indicates that this column has two non-zero elements, IROWT (1) = 1 and IROWT (2) = 3 indicate that these elements appear in row 1 and row 3 respectively and NCOLT (1) = 3 indicates that the significant column is column 3 of this transformation matrix.

The new values of the right-hand vector can now be obtained from equations 7.12 and 7.20 as:

$$\mathbf{b}' = \mathbf{B}^{-1}\mathbf{b} = \mathbf{T}_1\mathbf{b} = \begin{matrix} 4 \\ 5 \\ 1 \\ 7 \end{matrix}\begin{bmatrix} 5 \\ 20 \\ 15 \\ 15 \end{bmatrix}$$

The original elements of **b** can be overwritten by the corresponding values of **b**′. Also, the indices associated with the above vector indicate the variables that are now in the basis, these being at the end of the first iteration:

$$\mathbf{X}_b^{\mathrm{T}} = [x_4, x_5, x_1, x_7]$$

The actual multiplication by the transformation matrices can be performed using the method described in the previous chapter.

The above process can now be repeated for subsequent iterations. A summary of these iterations is as follows:

Iteration 2

$$\mathbf{C}_b = \overset{4}{[0,}\ \overset{5}{0,}\ \overset{1}{-16,}\ \overset{7}{0]}$$

$$\mathbf{C}'_a = \overset{1}{[*,}\ \overset{2}{-15,}\ \overset{3}{\circled{-16},}\ \overset{4}{*,}\ \overset{5}{*,}\ \overset{6}{16,}\ \overset{7}{*]}$$

$$\mathbf{P}_3 = \begin{array}{c} 4 \\ 5 \\ 6 \\ 7 \end{array}\begin{bmatrix} 0 \\ 0 \\ -1 \\ 1 \end{bmatrix} \qquad \mathbf{P}'_3 = \begin{array}{c} 4 \\ 5 \\ 1 \\ 7 \end{array}\begin{bmatrix} \textcircled{1} \\ 0 \\ -1 \\ 1 \end{bmatrix} \qquad \mathbf{T}_2 = \begin{array}{c} \\ 3 \\ 5 \\ 1 \\ 7 \end{array}\begin{array}{c} \begin{array}{cccc} 4 & 5 & 1 & 7 \end{array} \\ \left[\begin{array}{c:ccc} 1 & \cdot & \cdot & \cdot \\ 0 & 1 & \cdot & \cdot \\ 1 & \cdot & 1 & \cdot \\ -1 & \cdot & \cdot & 1 \end{array}\right] \end{array}$$

Location	1	2	3	4	5	6	7	8
VALUET	−1·0	1·0	1·0	1·0	−1·0	—	—	—
IROWT	1	3	1	3	4	—	—	—

Location	1	2	3	4
ICAPT	1	3		
NOZET	2	3		
NCOLT	3	1		

$$\mathbf{b}' = \mathbf{B}^{-1}\mathbf{b} = \mathbf{T}_2\mathbf{T}_1\mathbf{b} = \begin{array}{c} 3 \\ 5 \\ 1 \\ 7 \end{array}\begin{bmatrix} 5 \\ 20 \\ 20 \\ 10 \end{bmatrix} \qquad \mathbf{X}_b = \begin{array}{c} 3 \\ 5 \\ 1 \\ 7 \end{array}\begin{bmatrix} x_3 \\ x_5 \\ x_1 \\ x_7 \end{bmatrix}$$

Iteration 3

$$\mathbf{C}_b = \begin{array}{cccc} 3 & 5 & 1 & 7 \\ [0, & 0, & -16, & 0] \end{array}$$

$$\mathbf{C}'_a = \begin{array}{ccccccc} 1 & 2 & 3 & 4 & 5 & 6 & 7 \\ [*, & \textcircled{-15,} & * & 16, & *, & 0, & *] \end{array}$$

$$\mathbf{P}_2 = \begin{array}{c} 4 \\ 5 \\ 6 \\ 7 \end{array}\begin{bmatrix} 0 \\ 1 \\ 0 \\ 1 \end{bmatrix} \qquad \mathbf{P}'_2 = \begin{array}{c} 3 \\ 5 \\ 1 \\ 7 \end{array}\begin{bmatrix} 0 \\ 1 \\ 0 \\ \textcircled{1} \end{bmatrix} \qquad \mathbf{T}_3 = \begin{array}{c} \\ 3 \\ 5 \\ 1 \\ 2 \end{array}\begin{array}{c} \begin{array}{cccc} 3 & 5 & 1 & 7 \end{array} \\ \left[\begin{array}{ccc:c} 1 & \cdot & \cdot & 0 \\ \cdot & 1 & \cdot & -1 \\ \cdot & \cdot & 1 & 0 \\ \cdot & \cdot & \cdot & 1 \end{array}\right] \end{array}$$

Location	1	2	3	4	5	6	7	8
VALUET	−1·0	1·0	1·0	1·0	−1·0	−1·0	1·0	–
IROWT	1	3	1	3	4	2	4	–

Location	1	2	3	4
ICAPT	1	3	6	
NOZET	2	3	2	
NCOLT	3	1	4	

$$\mathbf{b}' = \mathbf{B}^{-1}\mathbf{b} = \mathbf{T}_3\mathbf{T}_2\mathbf{T}_1\mathbf{b} = \begin{matrix}3\\5\\1\\2\end{matrix}\begin{bmatrix}5\\10\\20\\10\end{bmatrix} \qquad \mathbf{X}_b = \begin{matrix}3\\5\\1\\2\end{matrix}\begin{bmatrix}x_3\\x_5\\x_1\\x_2\end{bmatrix}$$

Iteration 4

$$\mathbf{C}_b = \begin{matrix} 3 & 5 & 1 & 2 \\ [0, & 0, & -16, & -15] \end{matrix}$$

$$\mathbf{C}_a' = \begin{matrix} 1 & 2 & 3 & 4 & 5 & 6 & 7 \\ [*, & *, & *, & 1, & *, & 15, & 15] \end{matrix}$$

Since there are no negative coefficients in $\mathbf{C}_a'$ at this iterative step, the optimal solution has been obtained at the end of the third iteration. The basic variables at this point are therefore the solution to the problem, these being:

$$x_1 = 20, x_2 = 10, x_3 = 5, x_4 = 0, x_5 = 10, x_6 = 0, x_7 = 0.$$

The above numerical example gives the main features of a storage scheme and the arithmetic operations for the revised simplex method using the product form of the inverse. Many modifications of this basic scheme may be necessary for various reasons and these can be implemented. One possible modification is that reinversions may be needed within the solution. In this case, the storage scheme and program organization will depend on the method used for optimal pivot ordering. The ordering and factorization scheme described in chapter 6 would be suitable for this purpose.

7.3 Application to trans-shipment problems

7.3.1 The trans-shipment problem

Trans-shipment problems occur in many economic and engineering applications. These problems form a special type of linear programming problem in which the requirements and resources are expressed in terms of only one kind of unit and in which there are no restrictions on the number of routes between any origin and any destination.

When such problems are expressed in the standard linear programming form of $\mathbf{AX} = \mathbf{b}$, a coefficient matrix $\mathbf{A}$ has a simple and special structure consisting of elements which have numerical values of either +1, −1 or 0. This special structure is particularly suitable for sparsity programming and permits the solution to be achieved much faster than by either the simplex or revised simplex methods. This is achieved because any multiplications associated with this matrix involve only simple additions or subtractions of other quantities. Furthermore, because of the simple structure of $\mathbf{A}$, this matrix can be formulated as an incidence matrix. This property permits trans-shipment problems to be solved by the network approach, using the definitions and relations deduced for incidence matrices in Chapter 2. Because of these beneficial features, some problems which are not of the trans-shipment type are sometimes either approximated to it or partitioned to yield a part which is of that type. The partitioned problems can be solved by either the decomposition method of Dantzig and Wolfe or the partitioning method of Benders.

To illustrate the principles involved in trans-shipment problems and the application of the network approach in their solution, consider the following numerical example:

Minimize the transportation cost of 25 units of a certain commodity which is available from two sources, 0 and 3, and which is required at a destination, 2. The amount of commodity at each origin and the cost of transporting one unit of the commodity through the various routes is shown in matrix form below:

	destinations 0	1	2	3	availability constraints
sources 0	·	3	5	·	15
1	·	·	1	·	0
2	·	·	·	·	0
3	·	6	4	·	10
demand constraints	0	0	25	0	

The same problem may also be expressed in terms of a graph or network-type structure, as shown in Fig. 7.1. This can sometimes be more convenient and can indicate visually the structure of the trans-shipment problem.

The graph may be considered as a nodal network, where node 0 is arbitrarily chosen as the reference node. The supplies at the origins are represented by

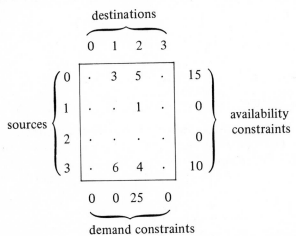

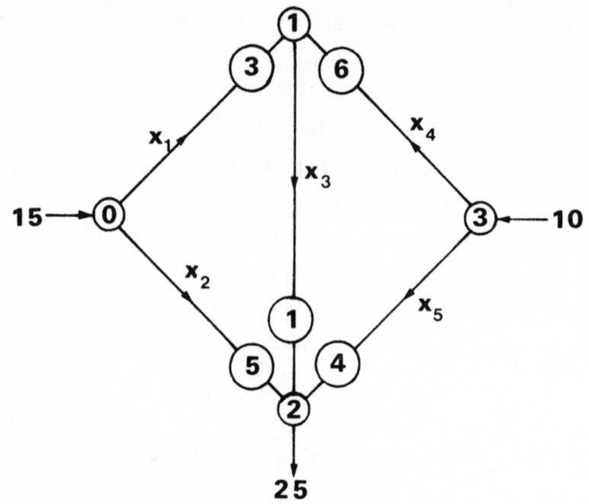

Fig. 7.1 Typical trans-shipment problem

positive flows into the nodes and the demands at the destinations by negative flows into the nodes. In addition, there are two elements associated with each branch. These are:

x_i = the amount of commodity being transported in the given direction along branch i

c_i = the cost of transporting a unit amount of commodity along branch i. For this example, these unit costs are shown in circles in Fig. 7.1.

From Kirchhoff's first law which states that the sum of flows entering a given node is equal to the sum of flows leaving that node, the following equations may be deduced:

$$x_1 + x_2 = 15 \tag{7.23a}$$

$$-x_1 + x_3 - x_4 = 0 \tag{7.23b}$$

$$-x_2 - x_3 - x_5 = -25 \tag{7.23c}$$

$$x_4 + x_5 = 10 \tag{7.23d}$$

For the reasons previously discussed in chapter 2, the equation relating to the reference node (equation 7.23a say) is omitted. The remaining equations (7.23b–7.23d) may be expressed as:

$$\begin{bmatrix} -1 & \cdot & 1 & -1 & \cdot \\ \cdot & -1 & -1 & \cdot & -1 \\ \cdot & \cdot & \cdot & 1 & 1 \end{bmatrix} \begin{bmatrix} x_1 \\ x_2 \\ x_3 \\ x_4 \\ x_5 \end{bmatrix} = \begin{bmatrix} 0 \\ -25 \\ 10 \end{bmatrix} \tag{7.24}$$

or in matrix form as:

$$\mathbf{AX} = \mathbf{b} \tag{7.25}$$

where **A** is the branch-nodal incidence matrix previously described and defined in Chapter 2.

The objective of this problem is to minimize the total cost, which can be expressed in the form:

$$\min Z = \mathbf{CX} \tag{7.26}$$

where, for the given numerical example:

$$\mathbf{C} = [3, 5, 1, 6, 4]$$

Also, since the flow in any given branch can only be in one direction, then:

$$\mathbf{X} > 0 \tag{7.27}$$

7.3.2 Basic feasible solution

The basic concept of trans-shipment problems is similar to that of the linear programming problems discussed previously. Both require optimization of an objective function. In the solution of a trans-shipment problem, therefore, there will be a set of basic variables $\mathbf{X}_b$ at each iterative step. Equation 7.25 can therefore be partitioned in terms of these variables to give:

$$[\mathbf{A}' \,\vdots\, \mathbf{B}]\begin{bmatrix}\mathbf{X}_a \\ \hdashline \mathbf{X}_b\end{bmatrix} = [\mathbf{b}] \tag{7.28}$$

Expanding equation 7.28 gives:

$$\mathbf{A}'\mathbf{X}_a + \mathbf{B}\mathbf{X}_b = \mathbf{b} \tag{7.29}$$

or

$$\mathbf{X}_b = \mathbf{B}^{-1}(\mathbf{b} - \mathbf{A}'\mathbf{X}_a) \tag{7.30}$$

where the matrix **B** is again a square non-singular matrix, defined as the basis matrix. In this case, its elements have numerical values equal to +1, −1 or 0. For the reasons concerning matrices of this type discussed in Chapter 2, this matrix corresponds to the branch-nodal incidence matrix of a tree of the network representing the trans-shipment problem. Therefore, there is a direct correspondence between the tree as defined in Chapter 2 and the basis as defined in section 7.2.

The basic variables $\mathbf{X}_b$ therefore correspond to the flows in the tree branches and the non-basic variables $\mathbf{X}_a$ correspond to the flows in the co-tree branches. The non-basic variables are set to zero, and therefore, from equation 7.30, the basic variables are given by:

$$\mathbf{X}_b = \mathbf{B}^{-1}\mathbf{b} \tag{7.31}$$

As shown in Chapter 2, the inverse of the tree branch-nodal incidence matrix is equal to the transpose of the branch-path incidence matrix. Therefore, instead of inverting the matrix **B** explicitly, the flows in the tree branches, $\mathbf{X}_b$, can be obtained by using the branch-path incidence matrix and tracing the supplies and demands from the origins and destinations to the reference node.

To illustrate this technique, consider the network shown in Fig. 7.1 and an initial tree deduced from it, as shown in Fig. 7.2. This tree corresponds to a basic feasible solution.

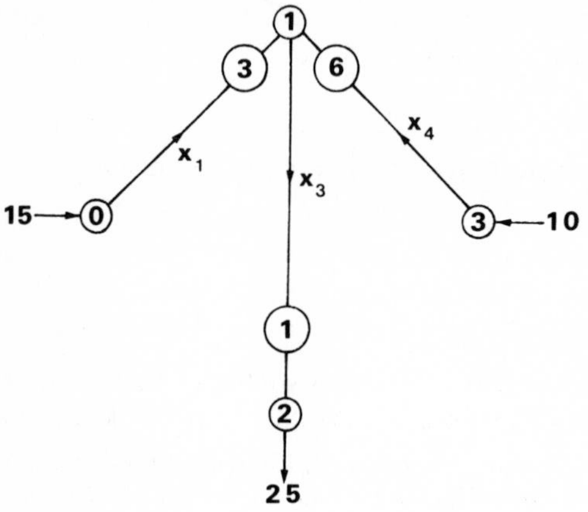

Fig. 7.2 Initial tree corresponding to an initial basic solution

Using Kirchhoff's law for node 2 of this tree, the flow in branch 3 is equal to the demand at node 2, that is:

$$x_3 = 25$$

Similarly, at node 3 and node 0

$$x_4 = 10$$

$$x_1 = 15$$

These values give an initial basic feasible solution.

Consider now the solution given by finding the paths from every independent node to the reference node and tracing the demands and supplies along these paths.

Branch 1 forms part of the paths from node 2 and node 3 to the reference node. Therefore, the flow in branch 1 is:

$$x_1 = -b_2 - b_3$$

The negative signs in this expression occur because the flow direction in branch 1 is opposite to the direction of the paths from node 2 and node 3 to the reference node.

Hence $\quad x_1 = -(-25) - 10 = 15$

Similarly

$$x_3 = -b_2 = 25$$

$$x_4 = \ b_3 = 10$$

It is evident, as should be expected, that both of the above methods give the same values for the basic variables x_1, x_3 and x_4. The comparison does, however, confirm that the branch-path technique enables the basic variables to be determined without inverting the basis matrix **B**.

7.3.3 Improving a basic feasible solution

The technique described in the previous section enables the basic variables corresponding to the initial tree branches to be found. The next step is to determine whether the solution given by these basic variables is optimal. If it is not optimal, it is necessary to find which non-basic variable should be brought into the basis, that is, which co-tree branch should be changed to a tree branch. As in the revised simplex method, the non-basic variable corresponding to the most negative coefficient c_i' is selected to become a basic variable. Although the principles are the same, the technique used to achieve this in trans-shipment problems differs to that previously discussed for the revised simplex method.

Consider equation 7.10 which was

$$\mathbf{C}_a' = \mathbf{C}_a - \mathbf{C}_b \mathbf{B}^{-1} \mathbf{A}' \qquad [7.10]$$

This can be expressed in partitioned matrix form as:

$$\mathbf{C}_a' = [\mathbf{C}_a \mid \mathbf{C}_b] \begin{bmatrix} \mathbf{U} \\ \hline -\mathbf{B}^{-1}\mathbf{A}' \end{bmatrix} \qquad (7.32)$$

Now **B** and **A**′ are the tree and co-tree branch nodal incidence matrices respectively. The relation between these matrices and the branch-loop incidence matrix **D**, given in chapter 2, is:

$$\mathbf{D}_t^T = -\mathbf{B}^{-1}\mathbf{A}' \qquad (7.33)$$

which gives:

$$\begin{bmatrix} \mathbf{U} \\ \hline -\mathbf{B}^{-1}\mathbf{A}' \end{bmatrix} = \mathbf{D} \qquad (7.34)$$

Hence, from equations 7.32 and 7.34,

$$\mathbf{C}_a' = \mathbf{C}\mathbf{D} \qquad (7.35)$$

From equation 7.35 and the definition of **D**, $\mathbf{C}_a'$ is the algebraic sum of the unit cost coefficients in the branches forming the basic loops. Using the initial tree shown in Fig. 7.2, the basic loops are as shown in Fig. 7.3. The dotted lines in this figure represent the corresponding co-tree branches.

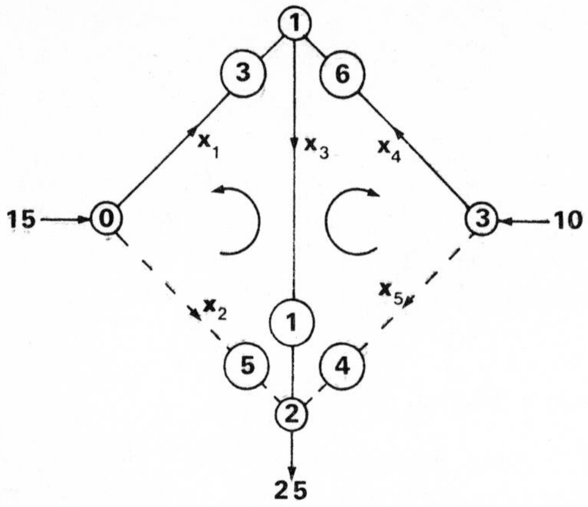

Fig. 7.3 Improving a basic feasible solution

The algebraic sum of the unit costs around the basic loop corresponding to co-tree branch 2 is:

$$c'_2 = c_2 - c_3 - c_1$$
$$= 5 - 1 - 3$$
$$= 1$$

The negative signs exist because the direction of the loop is opposite to the direction of branches 1 and 3.

Similarly, for the basic loop corresponding to co-tree branch 5:

$$c'_5 = c_5 - c_3 - c_4$$
$$= 4 - 1 - 6$$
$$= -3$$

The most negative coefficient c'_i is c'_5, indicating that co-tree branch 5 must be changed to a tree branch.

An alternative and computationally more efficient approach to the same problem is to first calculate the term $\mathbf{C}_b \mathbf{B}^{-1}$. Since matrix $\mathbf{B}^{-1}$ corresponds to the branch-path incidence matrix, then every column of $\mathbf{B}^{-1}$ represents the tree path from the corresponding node to the reference. Therefore every element of $\mathbf{C}_b \mathbf{B}^{-1}$ is equal to the algebraic sum of the unit costs of the branches along the tree path.

Let $\mathbf{V} = \mathbf{C}_b \mathbf{B}^{-1}$ (7.36)

Then for the above numerical example:

$$v_1 = -c_1 \qquad = -3$$

$$v_2 = -c_3 - c_1 = -1 - 3 = -4$$

$$v_3 = \ c_4 - c_1 = \ 6 - 3 = \ 3$$

Consider now a co-tree branch i with sending end node s and receiving end node r. The column $\mathbf{P_i}$ of $\mathbf{A'}$ corresponding to the co-tree i is:

$$\mathbf{P_i} = \begin{bmatrix} 0 \\ \vdots \\ 1 \\ 0 \\ \vdots \\ -1 \\ 0 \\ \vdots \end{bmatrix} \begin{matrix} \\ \\ s \\ \\ \\ r \\ \\ \\ \end{matrix}$$

Therefore for the variable c'_i, from equations 7.10 and 7.35:

$$c'_i = c_i - \mathbf{V}\mathbf{P_i} \tag{7.37}$$

or

$$= c_i - (v_s - v_r) \tag{7.38}$$

For co-tree branch 2 of the numerical example being considered, equation 7.38 becomes:

$$c'_2 = c_2 - (v_0 - v_2)$$

But v_0 is zero since node 0 is the reference node, and therefore there can be no branch between it and the reference.

Hence $c'_2 = 5 - [0 - (-4)]$

$= 1$

Similarly, for co-tree branch 5

$$c'_5 = c_5 - (v_3 - v_2)$$
$$= 4 - [3 - (-4)]$$
$$= -3$$

It is evident that the values of c'_2 and c'_5 obtained by this branch-path method are the same as those calculated using the previous branch-loop criterion. It also verifies that co-tree branch 5 must be changed to a tree branch, that is, the variable x_5 must be brought into the basis.

7.3.4 Interchanging basic and non-basic variables

When a co-tree branch i corresponding to a non-basic variable x_i is to be changed to a tree branch or basic variable, one of the tree branches must become a co-tree branch. In the revised simplex method, the appropriate variable was selected by first evaluating the column $\mathbf{P'_i}$ given by equation 7.23:

$$\mathbf{P'_i} = \mathbf{B}^{-1}\mathbf{P_i} \qquad [7.23]$$

where $\mathbf{P_i}$ is column i of matrix $\mathbf{A'}$.

The variable x_j leaving the basis was then determined using the criteria specified by equations 7.16 and 7.17.

It can be easily verified that for any given co-tree branch i of a trans-shipment problem, the elements p'_{ji} of column $\mathbf{P}'_i$ are:

$$p'_{ji} = \begin{cases} +1, & \text{if the direction of tree branch } j \text{ is opposite to the direction of the loop corresponding to co-tree branch } i, \\ -1, & \text{if the direction of tree branch } j \text{ is the same as the direction of the loop corresponding to co-tree branch } i \\ 0, & \text{if tree branch } j \text{ does not exist in the loop corresponding to co-tree branch } i \end{cases}$$

Therefore, using equations 7.16 and 7.17, the tree branch to be made a co-tree branch, that is the variable leaving the basis, is the one with the smallest flow and a direction which is opposite to the direction of the loop corresponding to co-tree branch i.

Using this criterion for the above numerical example, it can be seen from Fig. 7.3 that the tree branches to be considered are branches 3 and 4, since both of these have directions which are opposite to the loop corresponding to the co-tree branch to be made a tree branch. From section 7.3.2, these branches had flows of $x_3 = 25$ and $x_4 = 10$. Therefore branch 4 has the smallest flow and it must be made into a co-tree branch.

The same conclusions can be deduced from the following alternative argument. The criterion used in the revised simplex method is that for matrix **B** to remain non-singular as required, the vectors in the basis must remain linearly independent after bringing in a vector which was previously outside the basis. It follows therefore that the branch to go out of the basis must be one which, together with the branch brought into the basis, forms a loop: it is obvious that a tree, by definition, cannot include a loop.

Consider the network shown in Fig. 7.4 and assume that the co-tree branch between nodes 1 and 2, corresponding to a non-basic variable x_4, is to be brought into the basis. Also, define a loop direction which is the same as the direction of this co-tree branch.

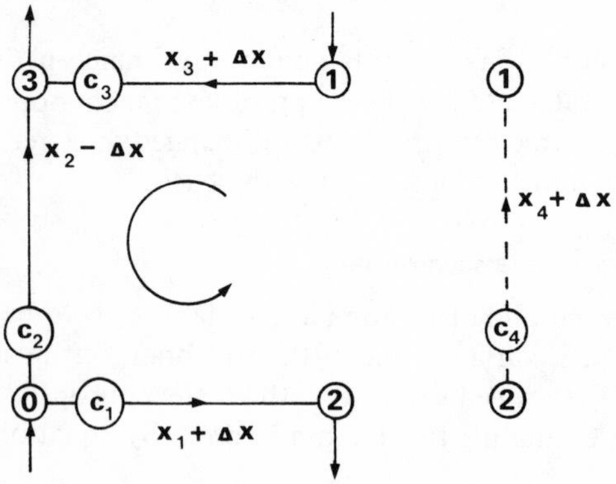

Fig. 7.4 Selecting a branch going out of a tree

One of the existing tree branches, say branch j, is removed from the tree and its flow becomes zero, i.e. $x_j = 0$. The flows in the loop will therefore change by an amount Δx, where $\Delta x = x_j$. The quantity Δx must be positive, since x_4 was zero, and equation 7.27 requires $\mathbf{X} \geqslant 0$. To achieve this positive quantity only those branches with directions opposite to the loop direction need be considered for removal. Furthermore, by choosing from these branches, the branch with the smallest flow, ensures that all the remaining flows cannot change direction and become negative; the remaining flows therefore stay feasible.

Consider now the application of this technique to the numerical example being analysed. The co-tree branch to be brought into the tree is branch 5. From Fig. 7.3 it is seen that the existing tree branches which have directions that are opposite to the loop direction corresponding to branch 5 are branches 3 and 4. The flows in these branches were found in section 7.3.2 to be:

$$x_3 = 25 \quad \text{and} \quad x_4 = 10.$$

The smallest flow is seen to be in branch 4 and therefore this branch is removed from the tree, that is variable x_4 is chosen to leave the basis. The new tree at the end of the first iteration is therefore as shown in Fig. 7.5.

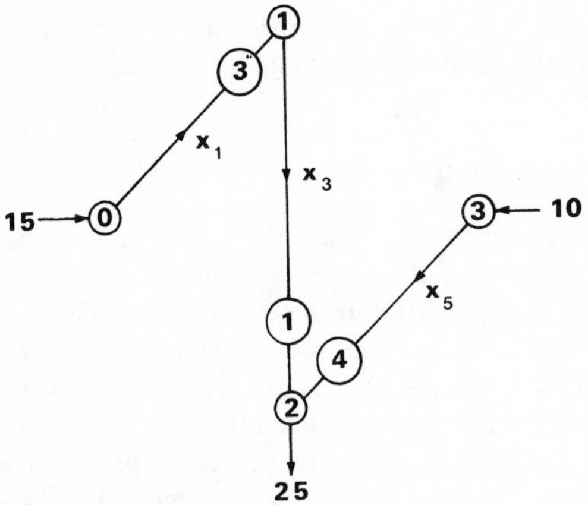

Fig. 7.5 Optimal solution

The reader can easily verify that the solution shown in Fig. 7.5 is optimal and:

$$x_1 = 15, \quad x_2 = 0, \quad x_3 = 15, \quad x_4 = 0 \quad \text{and } x_5 = 10$$

7.3.5 Steps of solution

The techniques discussed in the previous subsections for solving transshipment problems may look complicated compared to the revised simplex method for solving general linear-programming problems, discussed previously. The essence of the solution steps are, however, very simple, and in consequence the computational effort required is very much less than with the revised simplex method. The network approach to this type of problem is therefore powerful but requires skilful programming to be efficient.

The relative simplicity of the method is evident by considering the steps of solution that are summarized below:
(i) Find the initial tree and the flows in the tree branches.
(ii) Calculate the appropriate element of **V** for every independent node by summing algebraically the unit costs of transporting the commodity along the tree paths as expressed by equation 7.36:

$$\mathbf{V} = \mathbf{C}_b \mathbf{B}^{-1} \tag{7.36}$$

Hence calculate the elements of $\mathbf{C}'_a$ using equation 7.38:

$$c'_i = c_i - (v_s - v_r) \tag{7.38}$$

(iii) If all the elements of $\mathbf{C}'_a$ are greater than or equal to zero, the optimal solution is reached and the process is terminated. If one or more elements of $\mathbf{C}'_a$ are less than zero, the branch corresponding to the most negative element is selected to become a tree branch.
(iv) Trace the path along the existing tree from the receiving end of the co-tree branch selected in step (iii) to its sending end. From those branches having opposite directions to the traced path, select the one with the smallest flow to become a co-tree branch.
(v) Update the tree, find the flows in the new tree and repeat sequentially steps (ii) to (v) until the optimal solution is obtained.

7.3.6 Storage scheme and computational aspects

The solution time of the trans-shipment problem is greatly dependent on the time taken to trace computationally the paths between the reference node and the terminal nodes and between sending and receiving ends of the co-tree branches brought into the basis. This process of tracing forms steps (ii) and (iv) of the solution described in the previous section. These steps are logical and therefore their speed of execution depends on the storage organization. The matrices appertaining to these problems are generally sparse and are best stored in compact form. Efficient sparsity programming is therefore necessary in order to trace the required paths from the compacted data. One suitable storage scheme which permits easy access is described below.

The information concerning the routes of the trans-shipment problem may be stored in a table consisting of four arrays: VALUE (value of the flow), COST (cost of shipping a unit amount of the commodity), ISE (index of sending-end node number) and IRE (index of receiving-end node number). The table for the example shown in Fig. 7.1 is as follows:

Location	1	2	3	4	5
VALUE	15·0	0·0	25·0	10·0	0·0
COST	3·0	5·0	1·0	6·0	4·0
ISE	0	0	1	3	3
IRE	1	2	2	1	2

The locations in this table relate directly to the designated branch number, that is, location 1 refers to branch 1, etc. Also, the elements of the array VALUE in the above table are those for the initial tree and therefore represent the initial basic variables determined during step (i) of the solution. At successive iterative steps these values may change; the elements in the other three arrays, however, remain unchanged.

In certain high level programming languages, such as Fortran, zero cannot be used as a subscript to a variable. In such cases either the reference node must be designated by the next highest available node number or all the node numbers must be increased by one. Using this latter technique to re-number the network shown in Fig. 7.1 gives the network shown in Fig. 7.6.

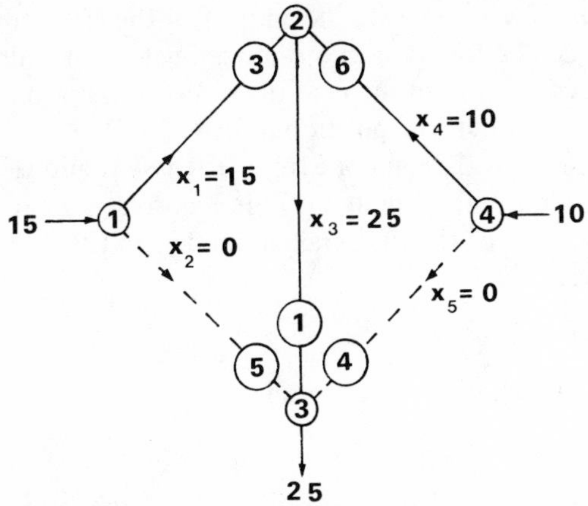

Fig. 7.6 Re-numbered trans-shipment problem

The storage table corresponding to the re-numbered nodes of the trans-shipment problem being considered is as shown below:

Location	1	2	3	4	5
VALUE	15·0	0·0	25·0	10·0	0·0
COST	3·0	5·0	1·0	6·0	4·0
ISE	1	1	2	4	4
IRE	2	3	3	2	3

In addition to this table, it is necessary to identify to which node each branch is connected. This is specified by the branch-incidence matrix which corresponds to the A-tableau of the general linear programming problem. It can be stored in a similar way using the arrays IBRANCH and INAP. The array IBRANCH is used to store the branch numbers connected to a node and the array INAP is used to store the index of node address pointer for the first branch number in IBRANCH which is connected to the corresponding node. A tree branch is indicated by a positive integer and a co-tree branch by a negative integer in the array IBRANCH.

For the network shown in Fig. 7.6, these two arrays are as shown below:

Location	1	2	3	4	5	6	7	8	9	10	11
IBRANCH	1	−2	1	3	4	−2	3	−5	4	−5	–

Location	1	2	3	4	5
INAP	1	3	6	9	11

The number of branches connected to each node could also be stored. However, the same information can be easily obtained from INAP without increasing the storage required. Using INAP, the address of the last branch connected to a given node can be found by subtracting unity from address pointer of the first branch connected to the next node. For this reason, the last entry in INAP indicates the first free position in IBRANCH.

To indicate the restructuring of the network from IBRANCH and INAP, find the branches that are connected to node 3. Node 3 corresponds to location 3 of INAP, and the location of the first branch number in IBRANCH connected to this node is given by:

INAP (3) = 6

The location of the last branch number is given by:

INAP (4) − 1 = 9 − 1 = 8

Therefore the relevant branch numbers are located in IBRANCH (6), IBRANCH (7) and IBRANCH (8), and the numbers are:

IBRANCH (6) = −2

IBRANCH (7) = 3

IBRANCH (8) = −5

This restructuring indicates that there are three branches connected to node 3, these being branches 2, 3 and 5. The information also indicates that branch 3 is a tree branch and branches 2 and 5 are co-tree branches.

It was stated at the beginning of this subsection that the solution time for the trans-shipment type of problem is greatly dependent on the time taken to trace the required paths of the tree. There are a number of ways in which this can be programmed using the information stored in the compact form discussed above. Most expert programmers will no doubt prefer to formulate the details of their own program using their own established techniques. It is, however, useful to discuss the general principles involved in the programming of tree tracing.

To do this let us consider the tracing of a path between the reference node and node 3, in order to determine the algebraic sum of the unit costs along this path and hence the value ν_3 at node 3. (It should be noted that, because of re-numbering, ν_3 refers to Fig. 7.6, and this is equivalent to the value ν_2 calculated previously for the network shown in Fig. 7.3.)

The process is commenced at the reference, that is node 1.

From INAP (1) = 1

and INAP (2) − 1 = 2,

the branches connected to the reference are stored in IBRANCH (1) and IBRANCH (2).

The element in IBRANCH (2) is negative, that is, it is a co-tree branch and can be ignored. Therefore, the only tree branch connected to the reference is:

IBRANCH (1) = 1

that is, branch 1.

The sending-end and receiving-end node numbers of branch 1 can be found in ISE (1) and IRE (1) respectively.

Now ISE (1) = 1

and IRE (1) = 2

Therefore branch 1 is connected between the reference and node 2, with a direction towards node 2.

The unit cost of shipping along branch 1 from the reference to node 2 is given in COST (1) and

COST (1) = +3

The above logic must now be applied to node 2, to find the tree branches connected to it and the unit costs involved in shipping from it to the next set of nodes. The reader will soon verify that the above logic shows that a tree branch (branch 3) exists between node 2 and node 3 and the unit cost of shipping from node 2 to node 3 along branch 3 is:

COST (3) = +1

Therefore the path between the reference and node 3 has been deduced, giving a unit cost of shipping from reference node to node 3 of:

$$\text{COST (1)} + \text{COST (3)} = 3 + 1 = 4$$

The required path is, however, opposite to the one so deduced, that is, we require the path from node 3 to the reference. Therefore, the unit cost of shipping from node 3 to the reference is:

$$-\text{COST (1)} - \text{COST (3)} = -4$$

$$\text{i.e. } \nu_3 = -4$$

which is the value previously calculated. In the same way, the other elements of **V** can be computed.

The other paths that have to be traced are those from the receiving end to the sending end of a co-tree branch being brought into the basis. This can be achieved using a variation of the above logic. A path can first be traced from the receiving end of the appropriate co-tree to the reference node, the node numbers along this path being stored. Secondly, a path is traced from the sending end towards the reference. This second path is terminated when a node is reached which also lies on the first path. Having found the node of intersection, the complete path between receiving end and sending end is determined. The branches forming this path can then be examined computationally and the appropriate one removed from the tree.

7.4 Other linear programming problems

The basic concepts of sparsity techniques and programming for two important categories of linear programming have been considered and described in the previous sections of this chapter. The authors believe that these considerations will greatly assist the reader in developing computationally efficient schemes and programs for his own particular type of problem. There are, however, several variations and adaptations of the general linear programming problems which are met in practice. These variations may include such aspects as upper and lower bounds to the solution of both the general linear programming problem and the trans-shipment problem. It may also be desirable in some cases to include decomposition, partitioning or diakoptical techniques to the problem. These variations can be included within the schemes proposed in this chapter by suitable minor modifications of the basic schemes. It is therefore hoped that the reader will find these sparsity techniques very useful in his programming exercises and the computational efficiency of them rewarding.

Bibliography

Bartels, R. H. & Golub, G. H.: The simplex method of linear programming using LU decomposition. *Commun. Ass. Computg Mach.* **12**, 266–8, 1969.

Beale, E. M. L.: Sparseness in linear programming. In *Large Sparse Sets of Linear Equations,* pp. 1–16. Academic Press, 1971.

Benders, J. F.: Partitioning procedures for solving mixed-variable programming problems. *Numl Math.* **4**, 238–52, 1962.

Brameller, A., Hindi, K. S. & Hamam, Y. M.: A network approach to the trans-shipment problem. *4th Iranian Conf. elect. Engng,* Iran, 1974.

Buchet, J. de: How to take into account the low density of matrices to design a mathematical programming package. In *Large Sparse Sets of Linear Equations,* pp. 211–18. Academic Press, 1971.

Carré, B. A.: An elimination method for minimal-cost network flow problems. In *Large Sparse Sets of Linear Equations,* pp. 191–210. Academic Press, 1971.

Dantzig, G. B.: *Linear Programming and Extensions.* Princeton University Press, 1963.

Dantzig, G. B.: Compact basis triangularisation for the simplex method. In *Recent Advances in Mathematical Programming,* pp. 125–32. McGraw-Hill, 1963.

Dantzig, G. B., Harvey, R. P., McKnight, R. D. & Smith, S. S.: Sparse matrix techniques in two mathematical programming codes. In *Sparse Matrix Proceedings,* pp. 85–99. IBM, 1969.

Dantzig, G. B. & Wolfe, P.: The decomposition algorithm for linear programs. *Econometrica* **29**, 767–78, 1961.

Forrest, J. J. H. & Tomlin, J. A.: Updating triangular factors of the basis to maintain sparsity in the product form simplex method. *Mathl Prog.* **2**, 263–8, 1972.

Gass, S.: *Linear Programming: Methods and Applications.* McGraw-Hill, 1958.

Graves, R. L. & Wolfe, P. (Ed): *Recent Advances in Mathematical Programming.* McGraw-Hill, 1963.

Hadley, G.: *Linear Programming.* Addison-Wesley, 1962.

Larson, L. J.: A modified inversion procedure for product form of inverse in linear programming codes. *Commun. Ass. Computg Mach.* **5**, 382–3, 1962.

Markowitz, H. M.: The elimination form of the inverse and its application to linear programming. *Managmt Sci.* **3**, 255–69, 1957.

Ogbuobiri, E. C.: Sparsity techniques in power system grid-expansion planning. In *Large Sparse Sets of Linear Equations,* pp. 219–30. Academic Press, 1971.

Orchard-Hays, W.: *Advanced Linear Programming Computing Techniques.* McGraw-Hill, 1968.

Orchard-Hays, W.: MP systems technology for large sparse matrices. In *Sparse Matrix Proceedings,* pp. 59–64, IBM, 1969.

Rabinowitz, P.: Applications of linear programming to numerical analysis. *Symp. appl. Math. Rev.* **10**, pp. 121–59, 1968.

Smith, D. M.: Data logistics for matrix inversion. In *Sparse Matrix Proceedings,* pp. 127–37. IBM, 1969.

Smith, D. M. & Orchard-Hays, W.: Computational efficiency in product form LP codes. In *Recent Advances in Mathematical Programming,* pp. 211–18. McGraw-Hill, 1963.

Tewarson, R. P.: On the product form of inverses of sparse matrices. *Symp. appl. Math. Rev.* **8**, pp. 336–42, 1966.

Tewarson, R. P.: On the product form of inverses of sparse matrices and graph theory. *Symp. appl. Math. Rev.* **9**, pp. 91–9, 1967.

Tomlin, J. A.: Modifying triangular factors of the basis in the simplex method. In *Sparse Matrices and their Applications,* pp. 77–85. Plenum Press, 1972.

Wolfe, P.: Trends in linear programming computations. In *Sparse Matrix Proceedings,* pp. 107–12. IBM, 1969.

Wolfe, P. & Cutler, L.: *Experiments in Linear Programming,* pp. 211–18. McGraw-Hill, 1963.

8
Application of Sparsity Techniques to Non-linear Problems

8.1 Introduction

In many engineering and non-engineering applications, the equations describing the problem to be solved are non-linear. Many aspects must be considered in general mathematical programming such as whether the problem is convex, or non-convex, has continuous or discrete (integer) variables and has static or variational cost functions. A discussion of these aspects is outside the scope of this book. It is sufficient to consider many non-linear problems which are characterized by a large, sparse network-type structure, for example power systems, only as continuous, convex and static problems. This reduces the optimization techniques to non-linear programming (NLP) in which the application of sparsity can be exploited as discussed in this chapter.

In general, these non-linear problems must be solved iteratively. One method by which this can be achieved is to linearize the equations at each iterative step and solve these linearized equations using the network-type of analysis.

One of the best methods for obtaining rapid convergence when solving a non-linear problem is to use the Newton (–Raphson) method. This method, however, can be computationally inefficient in terms of time and storage. Consequently other techniques have been dominant in the past. The more recent use and exploitation of sparsity techniques and programming in the solution of non-linear problems has allowed Newton's method to be used to great advantage and the method now becomes very attractive.

The application of Newton's method involves the solution of a set of Jacobian equations created from the first partial derivatives of the non-linear function being solved. The purpose of this chapter is to consider three typical types of problems that can occur frequently in practice, to discuss how and why the Jacobian matrix is created and to indicate how and where advantages can be gained by exploiting the sparsity structure of this matrix. The examples that are considered are power system load flow, general fluid (gas or water) flow problems and power system optimization. These examples may not directly relate to the reader's own problems, but many engineering and non-engineering problems have similar characteristics. The formulation of the problem and the solution techniques, though not necessarily the detailed mathematics, can therefore be very similar and the following discussion should be easily adaptable.

In these examples, the load flow problem is typical of any problem that can be formulated and solved using nodal analysis. This formulation is usually most advantageous when the non-linear equations have an exponent greater than unity. The general fluid flow problem is typical of those problems that are best formulated using the alternative loop methods, since nodal formulation would produce an exponent less than unity. In this latter case convergence would be obtained less rapidly and less reliably. With this type of problem the Jacobian matrix is best transformed to its nodal equivalent at each iterative step, to maximize the exploitation of sparsity. The third example of power system optimization is typical of optimization problems involving non-linear objectives and/or constraints.

8.2 Application to the load flow problem

8.2.1 Basic load flow equations

Load flow solutions are one of the most frequently performed digital computer calculations in power system analyses. For a considerable period of time they have been widely performed in power system planning, operation and control. With the advent of efficient computational algorithms, they are increasingly being performed in line outage security checks and optimal operation of very large systems. For greatest efficiency these modern algorithms rely on the use of sparsity techniques and programming.

The object of a load flow study is to determine the nodal voltages and hence the power flows in each branch of a transmission or distribution network for a given set of steady-state loading and operating conditions.

For a given electric network, Kirchhoff's Laws are used to form the nodal admittance equations:

$$\overline{\mathbf{Y}}\,\overline{\mathbf{V}} = \overline{\mathbf{I}} \tag{8.1}$$

where $\overline{\mathbf{I}}$ represents the equivalent nodal injected currents and is given by:

$$\overline{\mathbf{I}} = \frac{\overline{\mathbf{S}}^*}{\overline{\mathbf{V}}^*} \tag{8.2}$$

where $\overline{\mathbf{S}}$ is the net generation at a busbar (net nodal injected power) and is given by:

$$\overline{\mathbf{S}} = (\mathbf{P}_G - \mathbf{P}_L) + \mathrm{j}(\mathbf{Q}_G - \mathbf{Q}_L) = \mathbf{P}_N + \mathrm{j}\mathbf{Q}_N \tag{8.3}$$

where subscripts G, L and N represent generation, load and net quantities respectively.

Substituting equation 8.2 into equation 8.1 gives for node k:

$$\overline{S}_k - \overline{V}_k \Sigma \overline{Y}^*_{kj} \overline{V}^*_j = 0 \quad (8.4)$$

for j, $k = 1 \ldots n$

where n = number of nodes.

Equation 8.4 is non-linear and is therefore solved using iterative techniques. Consequently, at the first and intermediate iterations, the equation is not an equality and a power mismatch or residual will exist as given by:

$$\overline{\Delta S}_k = \overline{S}_k - \overline{V}_k \Sigma \overline{Y}^*_{kj} \overline{V}^*_j \quad (8.5)$$

for $j,k = 1 \ldots n$

where $\overline{\Delta S}_k$ will tend to zero at the final iteration.

Equation 8.5 can also be written in terms of polar co-ordinates. Using this representation and equating real and imaginary parts shows that the active and reactive power at any node k of a power system is:

$$\begin{aligned} \Delta P_{Nk} &= P_{Nk} - |V_k| \sum_{j \epsilon k} |V_j| (G_{kj} \cos \theta_{kj} + B_{kj} \sin \theta_{kj}) \\ \Delta Q_{Nk} &= Q_{Nk} - |V_k| \sum_{j \epsilon k} |V_j| (G_{kj} \sin \theta_{kj} - B_{kj} \cos \theta_{kj}) \end{aligned} \quad (8.6)$$

where ΔP_{Nk} and ΔQ_{Nk} are the power mismatches and tend to zero at the final iteration. These equations are real equations with real unknowns.

8.2.2 Solution of load flow problem

From equations 8.5 and 8.6, it is seen that four variables, P_{Nk}, Q_{Nk}, $|V_k|$ and θ_k are associated with each node. To solve these equations, two variables must be specified and then the two remaining variables can be calculated. The nodes can be classified into three types depending on which two variables are specified:

(i) slack-node – $|V|$, θ specified and P_N, Q_N unknown
(ii) P, Q-node – P_N, Q_N specified and $|V|$, θ unknown
(iii) P, V-node – P_N, $|V|$ specified and Q_N, θ unknown

In general, the equations corresponding to the slack-node can be removed from the set of equations given above. From the solution of the remaining equations, P_N and Q_N at the slack-node can be easily evaluated.

One of the most widely used methods for solving the load flow problem given by equation 8.6 is Newton's method. This has very rapid convergence properties. Before the use of sparsity techniques were introduced in the mid-1960s, however, its speed and storage requirements were unfavourable compared to the then-existing techniques of **Z**-matrix and **Y**-matrix iterative methods. This comparison is now reversed, but only if sparsity programming is used.

To illustrate how sparsity techniques and programming can be applied to the load flow problem, first consider how Newton's method is used to solve equation 8.6. Let **X** be the vector of all unknown variables and **U** the vector of all specified nodal variables. Equation 8.6 can then be solved by Newton's method as follows:

(i) select a number of equations from equation 8.6 equal to the number of unknowns in **X** to form a vector g:

$$g(\mathbf{X},\mathbf{U}) = 0 \tag{8.7}$$

(ii) use a Taylor's series expansion with higher order terms neglected; this linearizes the equations at each iterative step. Using initial estimates $\mathbf{X}^{(0)}$, solve for $\Delta\mathbf{X}$ from the following set of linear equations:

$$\mathbf{J}(\mathbf{X}^{(0)}, \mathbf{U})\,\Delta\mathbf{X} = -g(\mathbf{X}^{(0)}, \mathbf{U}) \tag{8.8}$$

where **J** is a Jacobian matrix having elements:

$$J_{jk} = \frac{\partial g_j}{\partial x_k}$$

(iii) from the solution of $\Delta\mathbf{X}$ and the initial estimates $\mathbf{X}^{(0)}$, determine new estimates for **X** from:

$$\mathbf{X}^{(1)} = \mathbf{X}^{(0)} + \Delta\mathbf{X} \tag{8.9}$$

This procedure can then be repeated until a sufficiently accurate solution has been obtained. This is generally achieved in 3 to 4 iterations for most practical problems and applications.

The Jacobian matrix has the same degree of sparsity as the admittance matrix **Y**. If this sparsity is exploited using the ordered elimination methods and compact storage schemes discussed in previous chapters for a set of simultaneous linear equations, Newton's method offers a very fast and reliable method for solving the non-linear load flow equations. Without using the sparsity property, the method can be computationally inefficient, particularly for large systems.

8.3 Application to general fluid flow problems

8.3.1 General method of solution

The types of fluid flow problems which occur in practice range from those that are simple and linear to those that are highly complex and non-linear. For water- or gas-distribution systems with interconnected pipes and a specified demand at various nodes, the problem is to evaluate the nodal pressures and the flows in the pipes. The solution to this problem may be obtained from a set of loop or nodal equations.

Since the equations are non-linear, the methods of solution are iterative, and therefore it is important to establish good initial values. These can be obtained easily in the case of the loop method from the definition of a tree. Also, experience and past investigations on fluid flow problems have established that methods based on loop formulation, such as the Hardy Cross method, are superior in convergence to nodal methods. It is therefore evident that the loop formulation can have a distinct advantage over nodal formulation in this type of iterative problem.

Many systems, including water- and gas-distribution networks, are characterized physically by their very high degree of sparsity. As discussed in Chapter 2, the sparsity of loop-formulated equations is dependent on the definition of the initial loops and it is very difficult to develop an efficient algorithm that can ensure maximum sparsity using these methods.

On the other hand, the nodal equations are very much easier to formulate and automatically give maximum sparsity. It is evident therefore that nodal formulation of the equations has the advantage of maintaining the original sparsity of the problem.

Using the highly efficient sparsity techniques developed in previous chapters for solving directly a set of large linear equations makes it possible to combine the distinct advantages of the two methods of formulation. The flow equations are first formulated using loop methods, then, using Newton's method, a well-conditioned Jacobian matrix and convenient starting values are produced. To retain and exploit the original sparsity of the problem, these Jacobian equations are transformed into nodal equivalents and solved using the techniques discussed previously. These techniques are developed and explained in the following subsections, using a gas-distribution system as an example.

8.3.2 The flow equations

Many equations exist which relate the flow to the pressure in a pipe. For low- and medium-pressure systems, the flow is assumed to be incompressible. A typical approximation used for the flow in a pipe p connected between nodes i and j is:

$$\Delta P_p = |P_i - P_j| = \frac{f^m{}_p}{k^m{}_p} \quad \text{(for low pressure)} \tag{8.10}$$

$$\Delta P_p = |P_i^2 - P_j^2| = \frac{f^m{}_p}{k^m{}_p} \quad \text{(for medium pressure)} \tag{8.11}$$

where P_i = pressure at node i

f_p = flow along pipe p between nodes i and j

$\dfrac{1}{k^m{}_p}$ = friction coefficient depending on pipe diameter and Reynold's number

m = constant; for low and medium pressure it is usually taken as $m = 2$.

Equations 8.10 and 8.11 can be written in general matrix form:

$$\Delta \mathbf{P} = \phi(\mathbf{f}) \tag{8.12}$$

where $\phi(\mathbf{f})$ = non-linear function of flow.

In their inverse form with $m = 2$, equations 8.10 and 8.11 are:

$$f_p = k_p \sqrt{|P_i - P_j|} \text{ sign } |P_i - P_j| \tag{8.13}$$

and $$f_p = k_p \sqrt{|P_i^2 - P_j^2|} \text{ sign } |P_i^2 - P_j^2| \tag{8.14}$$

8.3.3 Formulation of loop equations

The formulation of any network equation relies on the application of Kirchhoff's Laws. For the typical network shown in Fig. 8.1, the nodal equations are obtained by summing the flows at each node as follows:

$$\begin{aligned} f_4 - f_1 &= F_a \\ f_1 + f_3 - f_2 &= F_b \\ f_2 + f_5 &= F_c \end{aligned} \tag{8.15}$$

where F_i is the net load at node i and node 0 is taken as the reference node.

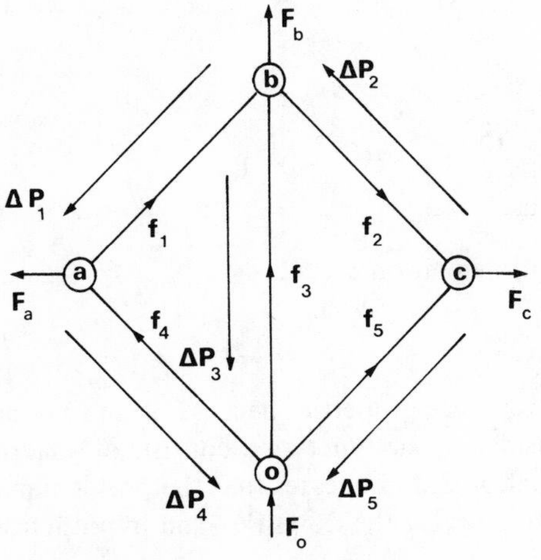

Fig. 8.1 Typical network for loop analysis

In general matrix form, equation 8.15 can be written (equation 2.10) as:

$$\mathbf{F} = \mathbf{C}\,\mathbf{f} \tag{8.16}$$

where $\mathbf{C}$ = branch-nodal incidence matrix.

Also, from Fig. 8.1, using Kirchhoff's second law:

$$\begin{aligned} \Delta P_1 + \Delta P_4 - \Delta P_3 &= 0 \\ \Delta P_2 + \Delta P_3 - \Delta P_5 &= 0 \end{aligned} \tag{8.17}$$

which can be stated in general matrix form as:

$$\mathbf{D}\,\Delta\mathbf{P} = 0 \tag{8.18}$$

where $\mathbf{D}$ = branch-loop incidence matrix.

Substituting equation 8.12 into equation 8.18 gives:

$$\mathbf{D}\,\phi(\mathbf{f}) = 0 \tag{8.19}$$

Defining tree and co-tree branches and partitioning equation 8.16 accordingly gives:

$$\mathbf{F} = [\mathbf{C}_t \mid \mathbf{C}_c] \begin{bmatrix} f_t \\ \hline f_c \end{bmatrix} \tag{8.20}$$

$$= \mathbf{C}_t\,\mathbf{f}_t + \mathbf{C}_c\,\mathbf{f}_c \tag{8.21}$$

Since $\mathbf{C}_t$ is square and non-singular, equation 8.21 can be expressed as:

$$\mathbf{f}_t = \mathbf{C}_t^{-1}\,\mathbf{F} - \mathbf{C}_t^{-1}\,\mathbf{C}_c\,\mathbf{f}_c \tag{8.22}$$

Now equation 2.3 shows that:

$$\mathbf{D}_t^{\mathrm{T}} = -\mathbf{C}_t^{-1}\,\mathbf{C}_c \tag{8.23}$$

Therefore, substituting equation 8.23 into 8.22 gives:

$$\mathbf{f}_t = \mathbf{C}_t^{-1}\,\mathbf{F} + \mathbf{D}_t^{\mathrm{T}}\,\mathbf{f}_c \tag{8.24}$$

and thus, from equation 8.24:

$$\mathbf{f} = \begin{bmatrix} \mathbf{f}_t \\ \hline \mathbf{f}_c \end{bmatrix} = \begin{bmatrix} \mathbf{C}_t^{-1}\,\mathbf{F} \\ \hline 0 \end{bmatrix} + \begin{bmatrix} \mathbf{D}_t^{\mathrm{T}}\,\mathbf{f}_c \\ \hline \mathbf{f}_c \end{bmatrix} = \mathbf{f}^{(0)} + \mathbf{D}^{\mathrm{T}}\,\mathbf{f}_c \tag{8.25}$$

Substituting equation 8.25 into equation 8.19 gives:

$$\mathbf{D}\,\phi(\mathbf{f}^{(0)} + \mathbf{D}^{\mathrm{T}}\,\mathbf{f}_c) = 0 \tag{8.26}$$

Equation 8.26 has good convergence properties because, in the gas- and water-flow problems being considered, the function ϕ consists of quadratic equations (8.10 and 8.11). Equation 8.26 indicates that the problem may be expressed in terms of initial estimates for the flows $\mathbf{f}^{(0)}$ and hypothetical loop flow corrections $\mathbf{D}^{\mathrm{T}}\,\mathbf{f}_c$.

By defining a tree for the problem of Fig. 8.1 as shown by full lines in Fig. 8.2 and applying Kirchhoff's second law, the equations at the r-th iterative step become:

$$\begin{aligned} \Delta P_1^{(r)} &= (f_1^{(0)} + q_1^{(r)})^2/k_1{}^2 + (f_4^{(0)} + q_1^{(r)})^2/k_4^2 - (f_3^{(0)} - q_1^{(r)} + q_2^{(r)})^2/k_3{}^2 \\ \Delta P_2^{(r)} &= (f_2^{(0)} + q_2^{(r)})^2/k_2{}^2 - (f_5^{(0)} - q_2^{(r)})^2/k_5{}^2 + (f_3^{(0)} - q_1^{(r)} + q_2^{(r)})^2/k_3{}^2 \end{aligned} \tag{8.27}$$

where the initial estimates $\mathbf{f}^{(0)}$ must satisfy the nodal flow constraints:

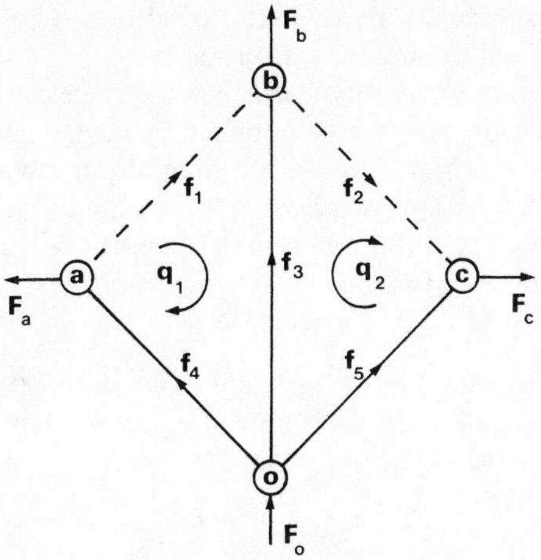

Fig. 8.2 Typical tree for determining loop flow correction

$$\begin{aligned} F_a &= f_4^{(0)} - f_4^{(0)} \\ F_b &= f_3^{(0)} + f_1^{(0)} - f_2^{(0)} \\ F_c &= f_2^{(0)} + f_5^{(0)} \end{aligned} \tag{8.28}$$

and where $\Delta P_l^{(r)}$ is the pressure mismatch or residual in loop l; these will tend to zero at the final iteration.

The solution for the unknown loop flows, $q_1 = f_1$ and $q_2 = f_2$, may be obtained at the r-th iterative step by using Newton's method from the equations:

$$\mathbf{J}\,\Delta\mathbf{Q}^{(r)} = -\ \psi(\mathbf{Q}^{(r)}) \tag{8.29}$$

where **J** is a Jacobian matrix having elements:

$$J_{jk} = \frac{\partial \psi_j}{\partial q_k}$$

From the solution of $\Delta\mathbf{Q}^{(r)}$ and the values $\mathbf{Q}^{(r)}$ determined at the r-th iterative step:

$$\mathbf{Q}^{(r+1)} = \mathbf{Q}^{(r)} + \Delta\mathbf{Q}^{(r)} \tag{8.30}$$

The solution of the problem can then be continued until the pressure mismatches come within acceptable limits.

8.3.4 Nodal solution of Jacobian equations

In the application of sparsity to the power system load flow problem, the Jacobian matrix had the same degree of sparsity as the network being considered. In this case, however, as discussed previously, the Jacobian matrix will have the same degree of sparsity as the original loop-formulated equations, which is likely to be less, and often considerably less, sparse than the network.

The pipe network equations are inherently better conditioned using the loop formulation and it is essential to preserve this property.

The maximum exploitation of sparsity and hence the achievement of minimum storage and computation time can be obtained by using nodal analysis at each iterative step. This is achieved by transforming the Jacobian equations (8.29) to equivalent nodal equations as follows:

The function ψ represents the pressure mismatches in the loops at the r-th iterative step and therefore equation 8.29 can be expressed as:

$$\mathbf{J}\ \Delta\mathbf{Q}^{(r)} = \Delta\mathbf{P}_l^{(r)} \tag{8.31}$$

Equation 8.31 which still describes the problem shown in Fig. 8.1 can now be represented by the equivalent network shown in Fig. 8.3, where

$$r_p = \frac{2q_p^{(r)}}{k_p^2}\ .$$

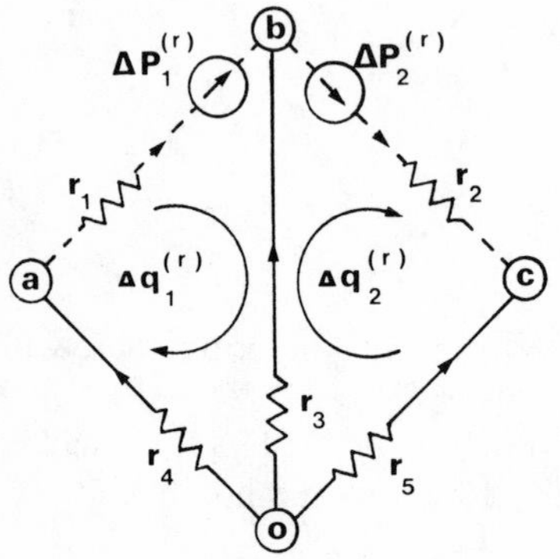

Fig. 8.3 Equivalent loop network of Jacobian equations

The equivalent resistances r_p in each branch of the network shown in Fig. 8.3 are the elements of the Jacobian matrix **J**; these being:

$$r_p = \frac{\partial}{\partial q}\left(\frac{q_p^2}{k_p^2}\right) = \frac{2q_p}{k_p^2} \tag{8.32}$$

The topology of the equivalent network shown in Fig. 8.3 is the same as the original network, and, in each loop l, the pressure difference source $\Delta P_l^{(r)}$ only exists in the co-tree branch. This equivalent loop network can easily be transformed into an equivalent nodal network by converting the pressure difference sources into nodal flows using Norton's theorem. This technique can be explained using the network shown in Fig. 8.4.

From Fig. 8.4b:

$$F_i = f_p' - f_p \tag{8.33}$$

and $F_j = f_p - f_p'$

$$\text{also } f_p' = \frac{P_i - P_j}{r_p} \tag{8.34}$$

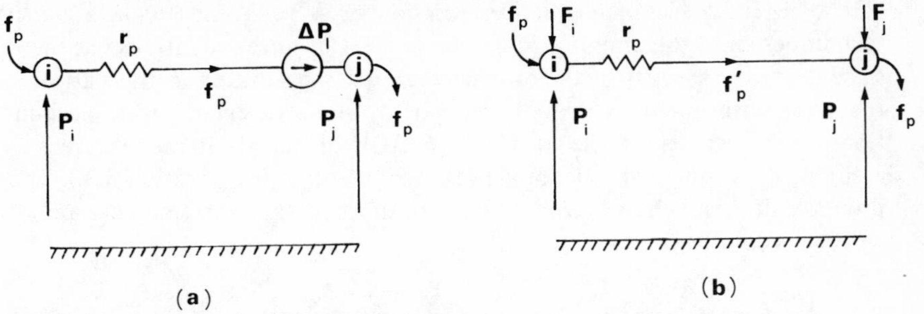

Fig. 8.4 Norton's theorem equivalent circuits
(*a*) branch with a series source; (*b*) equivalent branch with a nodal source

From Fig. 8.4a:

$$f_p = \frac{P_i - P_j + \Delta P_l}{r_p} \tag{8.35}$$

Substituting equations 8.34 and 8.35 into 8.33 gives:

$$F_i = -\frac{\Delta P_l}{r_p}$$

$$\text{and} \quad F_j = \frac{\Delta P_l}{r_p} \tag{8.36}$$

Equation 8.36 is the Norton's theorem transformation relation. Using these transformations, all the co-tree pressure difference sources ΔP_l can be converted to equivalent nodal flows F_i. The Jacobian equation (8.31) derived from the loop formulation can therefore be represented by an equivalent nodal network. For the problem being considered and shown in Fig. 8.1, the equivalent nodal network is shown in Fig. 8.5, where

$$y_p = \frac{1}{r_p}.$$

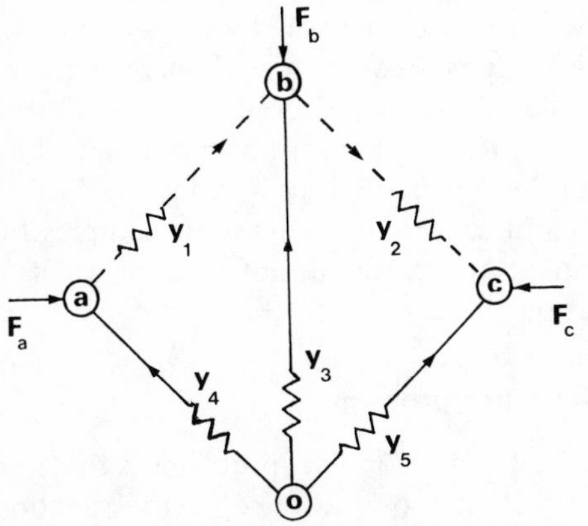

Fig. 8.5 Equivalent nodal network of Jacobian equations

The modified Jacobian equations can now be solved using standard sparsity techniques and programming. Since the new set of linear equations has the same degree of sparsity as the original network, the solution to this part of the iterative problem is achieved very quickly and efficiently. From the solution of these equations, the co-tree flows $\Delta Q^{(r)}$ at the r-th iterative step can easily be determined and the branch flows updated using equation 8.30. The process can then be repeated iteratively until the pressure mismatches are within acceptable limits.

8.4 Application to non-linear optimization problems

8.4.1 Non-linear optimization in power system analysis

Optimization using digital computers has become an extensive area of specialization and the procedures developed can serve as general-purpose tools in many applications. Many optimization problems involve linear objectives and linear constraints and can be solved very efficiently using the sparsity techniques discussed in previous chapters. In other cases, however, the objectives and constraints that are imposed by technical, physical, legal, political, financial requirements, etc. may be non-linear. To find the optimal solution in such cases, the equations defining the problem could be and often have been linearized and linear programming (LP) techniques used to obtain very fast approximate solutions. The errors introduced by this linear approximation can be excessive, however, for many present day requirements.

One application that requires non-linear techniques occurs in power system planning, in which the object is to ensure that future loads can be satisfied reliably and economically. The optimum size, location and time of installation of these new facilities can be planned effectively by use of non-linear optimization procedures.

Other applications occur in the operation of power systems, such as the determination of the optimal combination of system operating parameters including system security, system configuration, system operating cost, system losses, system voltage levels, etc. In most modern power systems, economical dispatch of loads is made in accordance with incremental cost procedures. These procedures take security constraints into account but without full concern for economy. With non-linear optimization techniques, however, the problems of economy and security against line overloads can be solved simultaneously. Although these techniques are more complex than the classical optimum dispatch method, their computational requirements are permissible for real-time operation.

8.4.2 Formulation of the optimization problem

The requirement of many practical and theoretical problems is to determine the optimum value of a scalar function known as the objective function. Some of the variables can be adjusted to certain values and these variables are known as control variables. In most problems the variables are connected or restricted by constraints due to the requirements discussed previously.

In the optimization of electrical power systems, many variations are possible. The problem in its basic and simplest form can, however, be described mathematically as follows:

Optimize the scalar function (objective)

$$\underset{\mathbf{U}}{\text{opt}}\ \mathbf{F}(\mathbf{X}, \mathbf{U}, \mathbf{P}) \tag{8.37}$$

and simultaneously satisfy the equations (constraints)

$$g(\mathbf{X}, \mathbf{U}, \mathbf{P}) = 0 \tag{8.38}$$

and/or the inequalities

$$h(\mathbf{X}, \mathbf{U}, \mathbf{P}) \leqslant 0 \tag{8.39}$$

where: **F** = objective function

X = vector of dependent variables

U = vector of independent (control) variables

P = vector of fixed parameters. For clarity, this vector will be omitted and assumed implicitly in the following equations.

The objective function **F**, which may be linear or non-linear, is to be minimized or maximized, depending on the problem, by the appropriate adjustment of the control variables **U**. The control parameters **U** may be the voltage magnitudes on **P**, **V**-nodes, transformer tap ratios, real power $\mathbf{P}_G$ available for economic dispatch, etc. The equalities given by equation 8.38 are the non-linear load-flow equation (8.7).

The inequalities given by equation 8.39 may represent, for example, the voltage limits on a **P**, **V**-node or a **P**, **Q**-node. These limits may be imposed on the control variables **U** and/or the dependent variables **X**. They are of the form:

$$\mathbf{U}_{min} \leqslant \mathbf{U} \leqslant \mathbf{U}_{max} \tag{8.40}$$

and $$\mathbf{X}_{min} \leqslant \mathbf{X} \leqslant \mathbf{X}_{max} \tag{8.41}$$

Control variables that have inequality constraints can easily be handled by setting its value to the appropriate limit whenever the solution exceeds that limit. Dependent variables can be handled by penalty methods which are added to the objective function **F**.

8.4.3 Solution of the optimization problem

One practical method for solving the optimization problem is the gradient (steepest descent) technique. Using the Lagrangian formulation, the problem can be stated as:

$$\mathbf{L}(\mathbf{X}, \mathbf{U}, \lambda) = \mathbf{F}(\mathbf{X}, \mathbf{U}) + \lambda^T\ \mathbf{G}(\mathbf{X}, \mathbf{U}) \tag{8.42}$$

where: λ = Lagrangian multipliers

F includes any penalty functions due to inequality limits on the dependent variables.

The conditions for the objective function **F** to be a stationary point are:

$$\frac{\partial L}{\partial U} = \frac{\partial F}{\partial U} + \frac{\partial G^T}{\partial U} \lambda = 0 \qquad (8.43)$$

$$\frac{\partial L}{\partial X} = \frac{\partial F}{\partial X} + \frac{\partial G^T}{\partial X} \lambda = 0 \qquad (8.44)$$

$$\frac{\partial L}{\partial \lambda} = G(X, U) \quad = 0 \qquad (8.45)$$

The gradient method can then be applied to the problem using the following steps:
(i) Assume a set of initial values for the control variables $\mathbf{U}^{(0)}$ and dependent variables $\mathbf{X}^{(0)}$. Use equation 8.45 and Newton's method to solve for $\Delta\mathbf{X}$ from the linear equations:

$$\mathbf{J}\,\Delta\mathbf{X} = -\,\mathbf{G}(\mathbf{X}^{(0)}, \mathbf{U}^{(0)}) \qquad (8.46)$$

where $\mathbf{J} = \frac{\partial \mathbf{G}}{\partial \mathbf{X}}$ is a Jacobian matrix having elements:

$$J_{jk} = \frac{\partial g_j}{\partial x_k}$$

From the solution of $\Delta\mathbf{X}$ and the initial estimates $\mathbf{X}^{(0)}$, determine new estimates for **X** from:

$$\mathbf{X}^{(1)} = \mathbf{X}^{(0)} + \Delta\mathbf{X} \qquad (8.47)$$

The Jacobian matrix **J** has the same degree of sparsity as the admittance matrix **Y**. Therefore, for the same reasons as discussed in the load flow application, exploitation of this sparsity structure permits a very fast and reliable solution to this part of the optimization problem.
(ii) The Lagrangian multipliers can now be found from the solution of equation:

$$\mathbf{J}^T \lambda = \frac{\partial \mathbf{F}}{\partial \mathbf{X}} \qquad (8.48)$$

The Jacobian matrix **J** has already been determined and factorized in order to solve equation 8.46. Therefore the solution of equation 8.48 is very simple and efficient if sparsity techniques are used.
(iii) The gradient indicating in which direction the maximum improvement to the objective function can be obtained is defined from calculus and equation 8.43 by:

$$\nabla U = \frac{\partial F}{\partial U} + \frac{\partial G^T}{\partial U} \lambda \qquad (8.49)$$

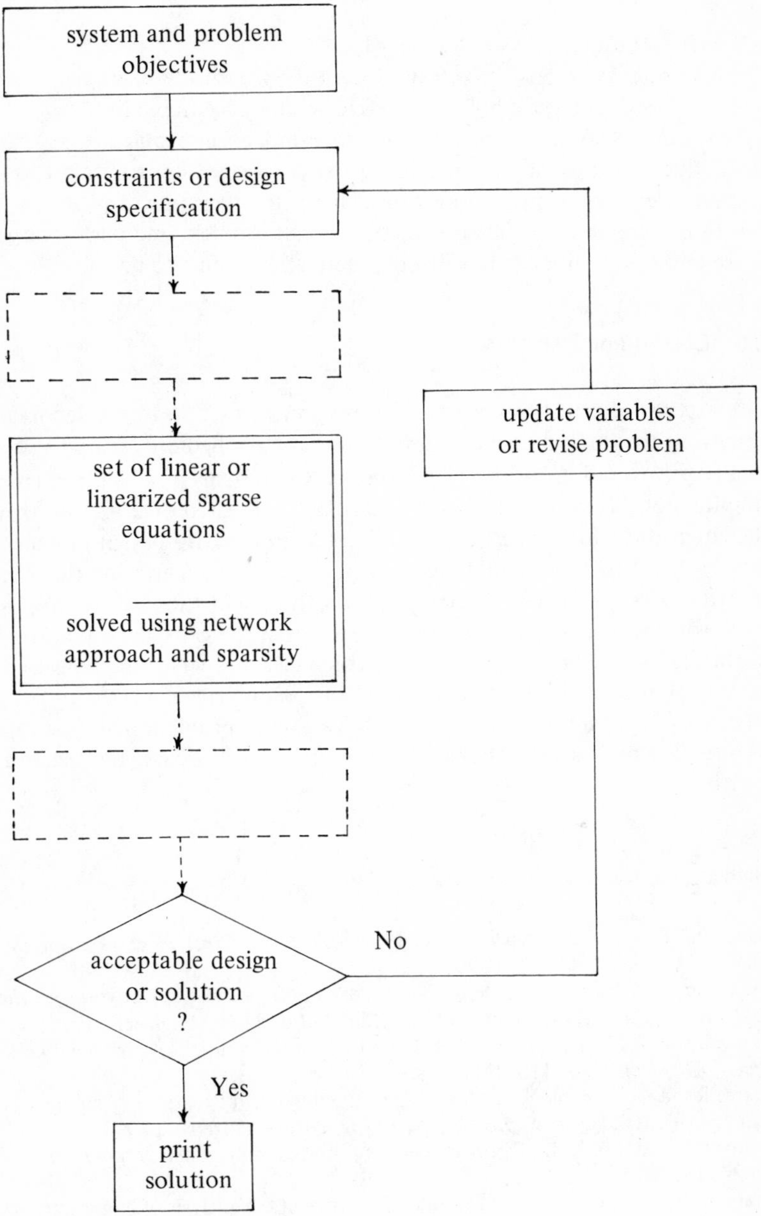

Fig. 8.6 Generalized flow chart of an iterative problem

(iv) A new set of control variables **U** can now be found from the initial estimates $\mathbf{U}^{(0)}$ and $\nabla\mathbf{U}$ by:

$$\mathbf{U}^{(1)} = \mathbf{U}^{(0)} + h\,\nabla\mathbf{U} \tag{8.50}$$

where h = a suitable scalar step length.

This process can now be repeated iteratively with the new variables $\mathbf{U}^{(1)}$ and $\mathbf{X}^{(1)}$ until equation 8.43 is satisfied within acceptable limits.

It is evident from this example of an optimization problem that sparsity techniques and programming can make an important improvement to the computational efficiency of part of a larger iterative type of problem. This can thus increase the efficiency of the overall problem and permit larger, more complex and more rewarding practical problems to be solved.

8.5 Concluding Remarks

In the types of problem discussed in the previous sections, the Jacobian equations to which sparsity techniques can be applied are created as a subset of a larger overall problem. There are many problems in which similar sets of simultaneous linear or linearized equations are created or used, for example, the equations which may be used in part of an iterative design project. Problems of this type can be illustrated typically by the generalized flow chart shown in Fig. 8.6, which clearly indicates the use of the network approach as a sub-routine of a larger problem. If, as is often the case, the equations relating to this part of the solution are sparse, then clearly sparsity techniques will be beneficial in their solution. The important point that an analyst or programmer should recognize is that, although his overall problem may not appear to be directly conducive to the application of sparsity, a particular part of it may be.

Bibliography

Baumann, R.: Some new aspects of load flow calculations. *Inst. elect. electron. Engrs Trans.* (PAS–85), **1965**, 1164–76.
Brameller, A. & Davenport, R.: *Optimum Pressure Control in Gas Networks for the Reduction of Losses*. Communication 882, Institution of Gas Engineers, 1972.
Brameller, A. & Hamam, Y. M.: Hybrid method for the solution of piping networks. *Proc. Instn elect. Engrs* **118**, 1607–12, 1971.
Brameller, A. & Hamam, Y. M.: Solution of the non-linear equations of piping networks by a hybrid method. *Int. Symp. System Engng*, Purdue University, 1972.
Brameller, A. & Lo, K. L.: Optimal operation of gas supply system. *Proc. Instn elect. Engrs* **118**, 1013–21, 1971.
Brammeler, A. & Allan, R. N.: The role of sparsity in the analysis of large systems. *Comput. Aided Des.* **6**, 159–68, 1974.
Brown, H. E.: *Solution of Large Networks by Matrix Methods.* John Wiley, 1975.
Broyden, C. G.: The convergence of an algorithm for solving sparse non-linear systems. *Math. Computn* **25**, 285–94, 1971.

Chang, A.: Application of sparse matrix methods in electric power system analysis. In *Sparse Matrix Proceedings,* pp. 113–21. IBM, 1969.
Dommel, H. W. & Tinney, W. F.: Optimal power flow solutions. *Trans. Inst. elect. electron. Engrs* (PAS–87), **1968**, 1866–76.
Schubert, L. K.: Modification of a quasi-Newton method for non-linear equations with sparse Jacobian. *Math. Computn* **25**, 27–30, 1970.

Index

acceleration, 53
    factor, 53

basic loop, 9
    linear transformation, 10
bi-factorisation, 71
branch, 5
    list, 6

coefficient matrix, 45
    asymmetrical, 45
    loop, 17, 23
    modifications, 40, 42
    nodal, 14, 22
computation, 125
    comparison, 128, 129
    fill in, 128, 129
    random networks, 127
    regular networks, 127
    storage, 128, 129
    times, 128, 129
connected graph, 5, 6
connection matrix, 6
convergence, 53
    acceleration, 53
    acceleration factor, 53
co tree, 7
    incidence matrix, 7, 8
    subgraph, 7
Crout elimination, 32

dependent node, 6
diagonal banding, 87
    bandwidth, 87
    comparison, 90
    major diagonal, 87
    merits, 95
    minor diagonal, 87, 88
    optimal, 96
diagonal matrix, 48
directed graph, 6
displacement method,
    simultaneous, 49, 50
    successive, 49, 50

factorisation, 58
    bi factorisation, 71
    comparison, 77
    LDH, 69
    LH, 64
    product form of inverse, 59
    triangular decomposition, 63
factor matrix, 59
    left hand, 71
    right hand, 71
fill in, 80, 128, 129
fluid flow, 161
    equations, 162
    equivalent network, 166
    Jacobian matrix, 165
    loop equations, 162, 163
    loop to nodal, 165
    mismatch, 165
    nodal equations, 162
    nodal solution, 165
    residuals, 165
    sparsity, 168

Gauss elimination, 28
Gauss–Jordan method, 35
Gauss–Seidel method, 47
graph, 5
    connected, 5, 7
    co tree, 7
    directed, 6

graph (*contd.*)
reduction, 80, 92, 93
sub, 5, 7
tree, 7
group relaxation, 51

ill conditioning, 53
incidence matrix, 5
branch loop, 9
branch nodal, 6, 145
branch path, 8, 146
co tree, 7, 8
tree, 7, 8
inversion, 35
program, 37
Gauss–Jordan method, 35

LDH factorisation, 69
LH factorisation, 64
linear equations, 4
choice of method, 55
comparison of methods, 56
direct methods, 28
iterative methods, 47
loop formulation, 16
nodal formulation, 11
relaxation methods, 51
solution of, 27
linear programming, 5, 132
basic feasible solution, 133
basic variables, 134
basis matrix, 134
changing variables, 134
non basic variables, 134
product form of inverse, 136
revised simplex method, 133
sparsity example, 138
steps of solution, 136
storage scheme, 136
load flow, 159
equations, 160
Jacobian matrix, 161
linearisation, 161
mismatch, 160
Newton Raphson, 161
residuals, 160
solution, 161
storage, 161

loop, 9
basic, 9
coefficient matrix, 17, 18, 23
impedance matrix, 17
linear transformation 10
lower diagonal matrix, 48

matrix
coefficient (*see* coefficient matrix)
connection, 6
diagonal, 48
factor, 59
factorisation (*see* factorisation)
incidence (*see* incidence matrix)
Jacobian, 158, 161, 165, 170
loop coefficient, 17
loop impedance, 17
lower diagonal, 48
lower triangular, 64
nodal admittance, 14
nodal coefficient, 14
transformation (*see* transformation matrix)
triangulation, 63
upper diagonal, 48

network changes, 40
modified matrix, 42
new solution, 42
network decomposition, 96
diakoptics, 97
matrix methods, 97
subsystems, 97
nodal coefficient matrix, 14, 22
node-dependent, 6
reference, 6
non linear equations, 4
solution, 19
non linear optimisation, 168
constraints, 168, 169
formulation, 168
Jacobian matrix, 170
Langrangian form, 169
Newton Raphson, 170
objectives, 168, 169
solution, 170
sparsity, 172
steepest descent method, 169

non linear programming, 158
Jacobian matrix, 158, 161, 165, 170
Newton Raphson method, 158, 161, 165, 170
types of problems, 159

ordering – banding schemes, 87
comparison, 94
dynamic, 86, 91
least connected branches, 86, 91
least new branches, 93
objectives, 83
optimal, 86, 94
pivotal, 83
pre, 86
principles, 83, 110
renumbering nodes, 85
semi optimal, 86

pivot, 28
choice of, 39
pivotal – element, 28
equation, 28
pivoting, 39
product form of inverse, 59
linear programming, 136
programming,
bi factorisation, 111
compacting, 125
dynamic ordering, 111
factorisation, 110, 116, 125
group elimination, 116
linear, 5
multiplication, 120
ordering, 110, 125
organisation, 124
pivoting, 111
simulation, 111, 125
solution, 122, 125
subroutines, 124
transformation matrix, 119

reference node, 6
relaxation methods, 51
residuals, 51

simultaneous displacement, 49, 50
sparsity – advantages, 100
coefficient, 21
directed elimination, 80
in fluid flow, 161
in linear programming, 132
in load flow, 159
in non linear optimisation, 168
in non linear programming, 158
in trans shipment, 142
loop coefficient matrix, 23
nodal coefficient matrix, 22
programming, 100
storage – adding new numbers, 102, 110
asymmetrical matrix, 105
compact form, 100
deleting elements, 110
double subscript method, 104
indexing, 100
linear programming, 136
linked list, 102, 107
list of numbers, 101
sparse matrix, 104
symmetrical matrix, 105, 106
transformation matrix, 119
subgraph, 5
connected, 7
co tree, 7
tree, 7
successive displacement, 49, 50

topology, 5
transformation,
nodal to loop, 19
transformation matrix, 59
left hand, 71
lower triangular, 65
product form, 59
right hand, 71
upper triangular, 67
trans shipment, 5
basic feasible solution, 145
changing variables, 149
graph representation, 144
incidence matrix, 145, 146
path tracing, 146, 154
steps of solution, 151
storage, 152
tree, 146

tree, 7, 8
  subgraph, 7
triangular,
  decomposition, 63
  lower matrix, 64
  upper matrix, 64
triangulation, 30, 63

upper diagonal matrix, 48
upper triangular form, 30

valency, 83